Seabed Minerals Series

Volume 3

Analysis of Processing Technology for Manganese Nodules

Seabed Minerals Series

Seabed Minerals Series

Volume 3

Analysis of Processing Technology for Manganese Nodules

**United Nations
Ocean Economics and Technology Branch**

Published in co-operation with the United Nations
by Graham & Trotman Limited

Published in 1986 by

Graham & Trotman Limited
Sterling House
66 Wilton Road
London SW1V 1DE

Graham & Trotman Inc.
13 Park Avenue
Gaithersburg
MD 20877, USA

in co-operation with the United Nations

© The United Nations, 1986

British Library Cataloguing in Publication Data

United Nations, *Ocean Economics and Technology Branch.*
Analysis of processing technology for manganese nodules.—(Seabed mineral series; v. 3)
1. Manganese—Metallurgy   2. Manganese nodules
I. Title   II. Series
622'.34.629   TN799.M3

ISBN 0-86010-349-8

ISBN 0 86010 349 8

D
622.3462
ANA

Typeset in Great Britain by Sprint, London EC2
Printed and bound in Great Britain by The Alden Press, Oxford

Contents

v

Preface to the Series

The deep seabed is one of the newest and potentially most reward-ing frontiers that has challenged mankind in its quest for knowl-edge and material achievement. Resources of the deep seabed promise to make an enormous contribution to the world's resource base if their potential is realized. At the present time, the resources of the deep seabed of immediate interest are in the form of manganese nodules which lie on the surface of the ocean floor and contain numerous metals — copper, nickel, cobalt, manganese, molybdenum, vanadium and titanium.

In addition to the potential for increasing the world's resource base, these minerals are particularly intriguing because they lie beyond the limits of national jurisdiction — they belong to no nation. According to a resolution adopted by the General Assembly of the United Nations, the area of the seabed and the resources therein are the common heritage of mankind.

While, for many years, manganese nodules have been the domain of marine geologists, chemists, and lawyers, interest is currently being shown by a broader audience — corporate and public entrepreneurs, policy-makers, managers, investors, engin-eers and technicians, as well as the international community of persons interested in the mineral and ocean industries, strategic minerals, and global development. The past two decades have witnessed the development of technology with a view to mining manganese nodules from a depth of nearly 15,000 feet and to extract the economically important metals from them. It has also witnessed the development, in the Third United Nations Conference on the Law of the Sea, of the legal and institutional framework in which the exploitation of the nodules will take place — who will exploit these resources, how this will be done, and who will benefit from this exploitation. At this moment, a water-shed appears to have been reached when decisions about application of technological prototypes of commercial operations are being taken and when the negotiations in the Law of the Sea

Conference culminated in the adoption of the United Nations Convention on the Law of the Sea. Thus, it seems to be an appropriate time to take stock, to assimilate widely diverse and dispersed information and analyses, and to make preparations for a future which envisages the resources of this newest frontier being utilized for the benefit of mankind as a whole.

It is in this context that the United Nations has undertaken the task of preparing a series of nine volumes examining different aspects of the development of manganese nodule resources.

The series starts with a discussion of how much of these resources exist in the world ocean. The quality and the quantity of the data collected through scientific expeditions and prospecting and exploration cruises are examined, the methods of estimating resources and reserves and their distribution on the basis of these data are explained, and a review of the currently available estimates is presented.

Since new technology had to be developed and tested for exploring, mining and processing manganese nodules, the next two volumes in the series describe and evaluate these technologies in terms of their efficiency and reliability.

Mine development is preceded by the delineation of a mine site. The fourth volume discusses the criteria used to designate a mine site. A potential mine site is defined as satisfying certain economic, geological and technological criteria; the size of the area may vary according to these criteria. This volume also develops and applies a simulation model which calculates the variations in the size of the mine site resulting from the variations in these criteria. Finally, based on available data and with the application of the simulation model, this volume attempts to identify areas meeting given sets of criteria.

The location of a processing plant for manganese nodules depends more on the availability and costs of complementary inputs, e.g. energy, chemicals, than on the traditional considerations of proximity to the mine or to the market. The criteria for selecting a site for a processing plant are examined in the fifth volume and a few likely sites of first generation processing plants are assessed according to these criteria.

The economics of a manganese nodule mining project is the crucial factor in the development of these resources. Currently, the economics can at best be based on order-or-magnitude estimates. These estimates are carefully examined and a financial profile of a potential seabed mining company is constructed in the sixth volume.

The seventh volume attempts to compare the economics of a manganese nodule mining project with alternative mining projects, e.g. nickel laterites.

The geological, technological and economic considerations for resource development can be translated to practice only within a defined regulatory framework. The eighth volume reviews the regulatory framework for the exploration and exploitation of seabed resources that is being devised by the international community. Finally, private companies, joined in several consortia, and public entities have already initiated and are carrying out work on mine development; the last volume of the series presents a summary of their activities and the status of their work.

The series is being prepared by the Ocean Economics and Technology Branch of the United Nations Department of International Economic and Social Affairs together with leading external experts. It draws and builds upon the information, data and expertise that the United Nations has been developing for over a decade concomitant with its work in the Law of the Sea Conference, and currently, in the Preparatory Commission for the International Sea-Bed Authority and for the International Tribunal for the Law of the Sea. These capabilities have been utilized in, and at the same time strengthened by, the development and maintenance at United National Headquarters of a data base on manganese nodules and a literature data base on marine minerals as well as the convening of a Group of Experts Meeting, the proceedings of which were published as *Manganese Nodules: Dimensions and Perspectives* (Dordrecht: Reidel, 1979).

The effort of the United Nations will succeed if the series can contribute to wider understanding of seabed resources and the problems and opportunities involved in their development, and can encourage constructive ideas and actions.

Acknowledgements

We gratefully acknowledge the contribution of Dr. Kwadwo Osseo-Asare of Pennsylvania State University's Department of Materials Science and Engineering in the preparation of this volume. We wish to thank Dr. J.C. Agarwal of Charles River Associates, Dr. Herbert Drechsler of H.D.D. Resource Consultants and Mr. D.W. Pasho of Energy Mines and Resources Canada for reviewing the manuscript and offering many helpful comments and suggestions.

Chapter 1

Introduction

The recovery of minerals from the world's oceans dates back to ancient times. The technology for extracting common salt by solar evaporation of sea water is known to have existed since 2290 B.C.[1] More recently and for about 100 years, offshore sand and gravel have been mined in the United Kingdom. The exploitation of offshore oil and gas now represents established technology in the petroleum industry. No metalliferous offshore resource has, however, generated as much interest and controversy as the deep sea polymetallic nodule resources (also known as manganese nodule resources). When first discovered in 1873 during the H.M.S. *Challenger*'s expeditions, limited analysis of their trace metal content revealed what was then considered insignificant amounts of copper, nickel and cobalt. A cobalt analysis of a nodule sample taken from the Tuamotu escarpment near Tahiti and performed in 1957 at the Scripps Institution of Oceanography in San Diego, California, sparked interest in nodules as potential mineral resources because the analysis revealed a cobalt content of greater than one percent. It was, however, not until John Mero's *Mineral Resources of the Sea* was published in 1965 that the quest to determine whether polymetallic nodules could be commercially exploited began in earnest.

Commercial interest in nodules has been primarily geared toward recovering the concentrations of nickel, copper, cobalt and manganese they contain. As in normal mining practice, this has meant that the search for nodule deposits has progressively been focussed on areas where the grades of the four metals in nodules, the abundance of such nodules and the topographical and geotechnical characteristics of the areas wherein they are found have converged to make them more attractive than other areas, in their potential to support commercial mining and processing operations.

Research undertaken so far in respect of establishing the commercial viability of mining nodules has included, *inter alia*, developing methods, techniques and equipment for exploring for, exploiting and processing nodules for their valuable metal content. The methods and technology for the first two stages have been the subject of the first two volumes of the series. These efforts, though invaluable, would, however, be of no commercial significance if the valuable metals contained in nodules can not be recovered (processed).

Polymetallic nodule resources contain a host of elements within them and a number of publications have listed up to 30 elements common to nodules.[2] Nodules usually contain between 25 and 30 percent (by weight) manganese, about 3 percent (combined weight) of nickel, copper and cobalt, approximately 30 percent (by weight) of a gangue fraction consisting of several clay minerals including montmorillinite, chlorite, kaolinite, and phillipsite, and biogenic fractions of calcium carbonate, silica, iron and from 30 to 40 percent water. No separate minerals of copper, nickel and cobalt have been found in nodules. Instead, the valuable metals are distributed in the manganese oxide phases: todorokite, birnessite and MnO. Moreover, and as a result of the above, nodules cannot be beneficiated by low cost physical means and their processing, where it entails drying, becomes a high cost component because of the amount of water they contain.

The selection of a process for the extraction of metals from polymetallic nodules is influenced not only by the usual technical and economic factors, but also by the choice of metals to be produced. The latter is controlled by processing costs, price and market demand. Manganese is the metal of principal value in the nodule ore, but is subject to severe competition from terrestrial ores. Nickel, on the other hand, is a superior target for recovery because of its historically rapidly growing market and the fact that new competition for nickel metal will be coming from lateritic sources that are at the present time, more expensive to process. Copper and cobalt from nodules face similar trade-offs in the market. Thus, a maximum flexibility in product mix is a desirable feature of the nodule metallurgical process.

The nodule metallurgical process had been the subject of study even before the four multinational consortia and the two groups of private companies and public agencies from France and Japan were formed. Inco Ltd.'s preconsortia efforts on the processing of polymetallic nodules date back to the late 1950s when early tests were conducted at their Copper Cliff facilities. During the early 1970s, the company supported processing studies on nodules at the University of California, Berkeley,[3] and conducted bench tests at their Port Colborne Laboratories.[4] By the time the consortia formed, Inco had apparently done at least preliminary work on thermal upgrading and smelting.[5] Preconsortia processing work by the DMOC companies appears limited to that done by the Sumitomo group of companies, principally Sumitomo Metal Mining Co. Ltd., (Sumitomo Kinzaku Kozon). As early as 1969, Sumitomo Metal Mining Co. Ltd., had completed bench-scale processing tests.[6] As early as 1970, Metallgesellschaft AG. of Germany had started their first laboratory processing tests on nodules and subsequently went on to work in co-operation with several other German companies experienced in extractive metallurgy. This "club for processing" investigated a variety of approaches and agreed to exchange data and results.[7] Some of the process routes were tested at pilot plant scale.[8]

Kennecott's early processing work on manganese nodules was in large part conducted by the research department of their Metal Mining Division in Salt Lake City, Utah. Beginning in 1962 and during the decade of the sixties, a variety of processing work, including literature reviews, contracted studies and in-house bench and small-scale pilot operations (smelting) were conducted. As part of their early work, experiments were conducted using nodules to remove sulphur compounds from gases.[9] At the University of Utah and, under contract to Kennecott, the Utah Engineering Experimental Station undertook benchscale testing of various process approaches as well as feasibility studies of smelting processes.[10] In the late 1960s, after the location of the site for processing studies was moved to Kennecott's Ledgemont Laboratory at Lexington, Massachusetts, work appears to have focussed on hydrometallurgical approaches, incorporating an ammoniacal leach of reduced nodules.

One of the most important conclusions generated as a result of the work described above, was that the metals contained in nodules could be extracted using, more or less, conventional methods such as the caron process (traditionally used to recover nickel and cobalt from laterites high in their content of iron) and smelting (traditionally used to recover nickel from lower iron and siliceous ores).[11]

At the present time, it may be said that a number of basic nodule metallurgical process routes have been established. This volume contains a general state-of-the-art description of what has been accomplished so far by synthesizing the relevant information available in the public domain and providing a bird's eye view of how the major technical and economic constraints have been circumvented. Its aims include providing the reader with a basic understanding of the factors that have led to the choice of metals to be produced, considerations in respect of the final product mix, competitive sources of metals from land-based minerals and experiences that have been drawn from the extractive metallurgy of land-based ores. With regard to the latter, lateritic nickel ores have been found to have a number of similarities with polymetallic nodules: first, in the actual metallurgical process and second as a source of competition for the commercialization of nodule ore.

In Chapter 2, a comparison of metal values within nodules and metal values within land-based deposits that are presently considered reserves is provided. This comparison is undertaken for the purposes of elaborating further upon the choice of metals to be produced from nodules and the interplay of market demand. Chapter 3 reviews the field of extractive metallurgy, concentrating on unit operations within this field and their applicability to nodule processing. Chapter 4 reviews the physical, chemical and mineralogical characteristics of polymetallic nodules providing background for the discussions in Chapter 5 that concentrate on the proposed metal extraction schemes for nodules. Chapter 6 summarizes the efforts undertaken by the various consortia in respect of the technology for processing nodules. Chapter 7 describes the conceptual flowsheets for first generation nodule processing plants, discusses the various flowsheets and also compares them with the processing routes for lateritic nickel ores. Chapter 8 sets forth some conclusions as to the proposed processing flowsheets and proceeds to foresee the requirements of development work needed before full-scale commercial exploitation.

NOTES

1. Mero, John L. Ocean mining — an historical perspective. *Marine Mining*, **1**, pp. 243–255.
2. Haynes, B.W., Stephen L. Law and D.C. Barron, 1982. *Mineralogical and elemental description of Pacific manganese nodules.* United States Bureau of Mines Information Circular 8906.
3. (a). Fuerstenau, D.W., *et al.*, 1969. *Metal Extraction from Ocean Manganese*

Nodules, The International Nickel Company of Canada, Ltd., (unpublished manuscript).

3. (b) Han, K.N., M. Hoover, and D.W. Fuerstenau, 1974. Ammonia-Ammonium Leaching of Deepsea Manganese Nodules. *International Journal of Mineral Processing*, **1**, pp. 215–230.

3. (c) Hoover, M., K.N. Han and D.W. Fuerstenau, 1975. Segregation roasting of Nickel, Copper and Cobalt from Deepsea Manganese Nodules, *International Journal of Mineral Processing*, **12**, No. 2, pp. 173–185.

4. *Metals Week*, 8 January, 1973.

5. Sridhar, R., 1974. Thermal Upgrading of sea nodules. Paper presented at the 103rd Annual AIME Meeting, Dallas, Texas.

6. Sumitomo Shoji Kaisha, Ltd., 1972. An historical review of manganese nodule development by the Sumitomo Group, (unpublished manuscript), 15 Sept. 1972, 4 pp.

7. Meyer-Galow, E., U. Boin and K.H. Schwarz, 1973. The marine nodules project: Spotlights from the view of process engineering. In Morgenstein, M., ed., *Papers on the Origin and Distribution of Manganese Nodules in the Pacific and Prospects for Exploration*. Honolulu, Hawaii, 23–25 July, 1973, pp. 131–138.

8. Kauczor, H.W., H. Jaunghanb and W. Roever, 1973. The hydrometallurgy of metalliferous solutions in the processing of manganese nodules. *Inter-Ocean 1973*, Dusseldorf, Germany, 13–18 Nov., pp. 469–473.

9. Zimmerley, S.R., *Use of Deep Sea Nodules for Removing Sulphur Compounds from Gases* U.S. Pat. 3,330,096, 11 July, 1967, 5 pp.

10. Hanson, C.K., G. Ramadorai and S.W. Wu, 1968. *Metals recovery from manganese sea nodules*. Interim report, Bear Creek Mining Company, 22 pp.
Beck, R.R. and M.E. Messner, 1970. Copper, nickel, cobalt and molybdenum recovery from deep sea nodules. In Ehrlich, R.P., ed., *Copper Metallurgy*. Metallurgical Society of the AIME, New York, pp. 70–83.

11. O'Kane, P.T., 1979. Energy consumption and economic trends in the production of nickel from laterites. In Evans, D.J.I., Shoemaker, R.S. and Veltman, H., eds., *International Laterite Symposium,* Society of Mining Engineers of the AIME, New York, pp. 503–523.

Chapter 2

Issues Associated with Processing Manganese Nodules

MANGANESE NODULES AS MINERAL RESOURCES

Mineral resources comprise all types of rocks which can be made to yield useful metals or industrial minerals. For normal commercial considerations, two conditions need to be added to the above statement:

(i) It must be technically feasible to extract the useful products at costs within the limits imposed by their commercial value; and,

(ii) the size of a deposit and its valuable metal content must be large enough to justify the capital investment needed for its exploitation as well as to sustain associated developments and commercial commitments for an appropriate period.

The manufacturing process whereby a raw material (rock or ore) is converted into a refined metallic finished product is referred to as the extraction of metals. The various steps, or unit operations, that form the tie linking the raw ore to the refined metal are characterized by their ability to effect various specialized kinds of separations. In the initial stages, the separations tend to involve primarily physical techniques and the various operations fall under the discipline of mineral processing.[1] In the later stages, separations are largely dependent on the application of chemical techniques and the corresponding processes constitute the discipline of chemical or extractive metallurgy.[2]

In the evaluation of a mineral property therefore, the mineral processing engineer is a key member of the evaluation team because he can provide a critical answer to the question, "can this ore be processed economically?" The early estimation of the processing approach is very important because, for many metalliferous ores, the capital investment for the mill, smelter, water supply and tailings disposal exceeds the cost of the mine. In general, in mineral processing, both capital and operating costs escalate very rapidly in the following order: gravity concentration of alluvial ores, gravity concentration of lode ores, flotation, simple hydrometallurgical processes and complicated hydrometallurgical and pyrometallurgical processes.[3] The latter two processes may cost 100 to 1000 times more per ton treated than the processing of a simple alluvial deposit.

In respect of processing nodules therefore, their physical properties are an essential consideration in the selection of the proper extraction process. The nodule matrix, in contrast to sulphide minerals, is a mixture of iron and manganese oxides which cannot be easily beneficiated and does not contain sulphur which contributes considerable energy savings (autogenous) during smelting. The major metal values, nickel, cobalt and copper, are not present as separate minerals, but are distributed throughout the oxide matrix. Nodules contain fine pores (100 Å), high porosities (60% volume), and a considerable amount of water (45–50% weight). The valuable metals are most readily recovered by reduction of the tetravalent manganese oxide matrix to manganous oxide.

Table 1. Average Element Concentrations (wt. %) in Nodules from the Central Indian Basin as Compared to an area Southeast of Hawaii

Metal	Central Indian Basin[a]	Area Southeast of Hawaii[b]
Mn	26.1 ± 5.7	24.8 ± 4.3
Fe	7.6 ± 3.3	6.4 ± 2.4
Co	0.12 ± 0.05	0.20 ± 0.09
Ni	1.20 ± 0.35	1.30 ± 0.30
Cu	1.16 ± 0.43	1.21 ± 0.40
Ni + Cu + Co	2.49 ± 0.71	2.71 ± 0.64
Cu/Ni	0.98 ± 0.28	0.93 ± 0.20
Mn/Fe	4.2 ± 2.0	4.4 ± 1.8

[a] Based on 25 assays from 15 locations
[b] Based on 214 assays
Source: Manganese nodule resources in the Indian Ocean, 1980
    J.G. Frazer and L.L. Wilson, *Marine Mining*, **2**, no. 3.

Table 1 illustrates the range of element concentrations found in nodules in two areas in which a considerable amount of prospecting for nodules has taken place. These areas are in the Central Indian Basin and in an area southeast of Hawaii. In reviewing Table 1 it can be seen that for the first generation of nodule mining, typical metal values will range from about 1–2% nickel, 1–2% copper, 0.1–0.3% cobalt, and 25–35% manganese on a dry weight basis. The sizes of mined nodules will of course vary depending on the mining method used. The material delivered for processing may also vary from a finely comminuted sludge to whole nodule chunks. Finally, the processing plant must be designed to deal with the problems of material handling as well as the presence of a substantial amount of sea water.

For many years, there has been a debate in the nodule mining literature about the pyrometallurgical versus the hydrometallurgical processing routes for nodules. One consortium championed each process, and it appeared that there was an almost mystical significance in the choice of a particular alternative. The choice of a processing route depends on a number of historical, corporate, economic and technical factors, which change over a period of time. It should be remembered that over the years much analysis has been devoted to the nodule mining industry with no history of proven production. To date, no one has even processed nodules continuously on a pilot plant scale; therefore, there is nothing particularly sacred about the pronouncements of a consortium that is wedded to one processing route. As the underlying decision variables and objectives change, as they must, the chosen processing route might also change.

The choice of a processing route will also be affected by the characteristics of the companies which are members of the consortium. If it is a mining company that is charged with developing the processing technology for a consortium, it stands to reason that the mining company would prefer an alternative which is similar to processes it uses on other ores, e.g. laterites. There would be economies in research and development, equipment development and design, and in personnel costs. The personalities and experience of research directors often play a significant role in the direction taken by consortia in developing process alternatives. Once a particular individual leaves a company, therefore, the company's research could take a new direction. If a consortium is headed by a non-mining company, then that company might simply select the best process technology on the market for its purposes and simply contract out the processing work. The role of corporate and human factors in setting directions for process development should be kept in mind.

In considering the development of a process, perhaps the first question is "what metals are to be produced from nodules and in what form?" The choice is conditioned by the relative cost of producing a product from nodules versus the competitive terrestrial ore. An evaluation must be made of the competitiveness of metal production from nodules; consideration must be given to the size of the market for the product in question (e.g. low-grade ferromanganese) and to the future prospects for the market, including demand and the evolution of price. To be consistent with the literature, it is assumed that nickel, copper and cobalt will be produced from manganese nodules. The question of whether or not manganese will be produced is still problematical.

Metal markets are traditionally volatile, and they tend to make forecasters look consistently bad over time; however the initial step in evaluating any mineral resource is to determine the value of the contained metals assuming complete recovery, as shown for a typical nodule ore deposit in Table 2. In this table, the long-term price forecasts for manganese, nickel, copper and cobalt are taken to be 95c/lb., $3.90/lb., $1.25/lb., and $3.00/lb. respectively. From Table 2, manganese immediately stands out as the major component in terms of value followed by nickel, copper and cobalt.

Table 2. Metal values per day short ton of ore

Metal	Grade %	lbs/ton	Price ($/lb)	Value $	% of total value
Manganese	27.0	540	0.95[a]	513	78.4
Nickel	1.3	26	3.90	101.4	15.5
Copper	1.1	22	1.25	27.5	4.2
Cobalt	0.2	4	3.00	12.0	1.8
				654	100.0[b]

[a] Price in ferromanganese.
[b] Rounded off.

Recovery of valuable trace elements in nodules, e.g. molybdenum and vanadium, is not often explicitly covered in the processing literature. A sufficient price incentive relative to cost would lead metallurgists to look into ways of incorporating their recovery onto a chosen process. Since the industry has not yet arrived at proving commercial scale processing capability, the choice of metals to be produced from nodules is not irrevocably fixed and therefore by implication, neither is the processing route.

Finally, there are the direct costs associated with the choice of a particular processing alternative. These are the capital costs for plant and equipment, and the operating costs, particularly energy costs, to which mining and processing operations are very sensitive. Additionally, there are costs associated with meeting certain pollution and safety standards, which are site-specific, i.e. they depend on where exactly the plant is to be located. It would be premature to discuss these, however, so environmental costs will be ignored in this analysis

The second step in evaluating a mineral resource, is to consider and take into account market demand. Markets may change even after this consideration; therefore, a process with maximum flexibility in the selection of metal and product forms is highly desirable. A short summary of the existing and forecasted markets for the four metals in the long-term is presented. Following this, a brief analysis of implications in respect of priority metals is undertaken.

MANGANESE

Metallurgical uses of manganese in iron and steel production accounts for about 97% of all manganese ore consumption. The remaining uses are in dry cells and chemical applications. Apart from the direct use of manganese ore in blast furnaces producing pig iron, manganese for steel-making is used in three forms:

    (i)   ferro-manganese (with a manganese content of 78%)
  (ii)   silico-manganese (with a manganese content of about 65% to 75%); and
 (iii)   spiegeleisen (with a manganese content of about 16% to 30%).

Worldwide, the average quantity of manganese used per ton of steel has remained fairly constant despite technical innovations such as external desulphurization, continuous casting and the basic oxygen process, all aimed at reducing manganese requirements.[4] The demand for manganese depends, therefore, on the demand for steel, and the rates at which manganese consumption has expanded in various countries reflect the growth rates of steel production.

Industrialized countries consume about 50% of the world's output of manganese. Their demand has grown at an average annual rate of 3.0% between 1960 and 1977. Japan, for example, starting at a very low level and with rapidly growing steel production, had the fastest expansion of manganese consumption, an average annual rate of 10.1% a year, during the period. In the long term, manganese consumption is projected to expand at an average rate of 2.7% a year during the period 1975 to 2000. Developing countries as a group, consume only 11% of the world's output of manganese. Their consumption has grown rapidly during the past 25 years, mainly as a result of the expansion of their domestic steel production capacities, the establishment of local iron and steel industries and dry cell battery plants. Centrally planned economies account for 39% of world manganese consumption, which is met mostly by supplies from the vast reserves in the USSR.

World production of manganese ore has been on the order of 29 million tons per year with a manganese content of 9 to 10 million tons. Table 3 summarizes data on global production and reserves of manganese.

There are ample manganese ore reserves and supplies. The demand for manganese ore is highly price inelastic, hence price fluctuations largely reflect changes in supply. The long-term price trend is projected at being largely affected by the excess availability of manganese ore which is also expected to have a higher metal content. This higher metal content is expected to partially offset projected higher mining costs. Large scale capital intensive technology is also expected to lead to lower unit costs of production. Therefore, prices in real terms are projected to decline gradually in the long term.

Analysis of Processing Technology

Table 3. World Production and Reserves of Manganese

From: Lévy, J. -P. and N. A. Odunton, 1984. Economic impact of seabed mineral resources development in light of the Convention on the Law of the Sea, *Nat. Res. Forum*, **8**, no. 2.

Country	World manganese production (1980)[a]		Mine production capacity (1985) contained metal (10³ tons)[b]	Estimated reserves (1983) contained metal (10³ metric tons)[c]
	Per cent	Gross weight (10³ tons)		
USSR	35	10748	4200	2508024
South Africa	30–48 +	6278	2900	2630837
Brazil	38–50	2601	1350	62595
Gabon	50–53	2366	1500	172365
Australia	37–53	1872	1500	195952
India	10–54	1814	700	39916
China	20 +	1750	500	29030
Mexico	35 +	493	300	7076
Ghana	30–50	278	150	6000
Morocco	50–53	145	30	1542
Hungary	30–33	97	40	—
Japan	24–28	88	30	—
(A) Total		28530	13200	3653324
(B) World Total		29091	13370	3653324
(A) as per cent of (B)			98.7	100

Sources: [a] USBM, 1981.
      [b] USBM, 1980.
      [c] Jones, 1983.

[a]Production is not quoted in contained metal. It is quoted in the range of manganese ore grades and total tonnages mined in 1980.

NICKEL

In land-based deposits, nickel occurs in two general types of ore, sulphides and oxides (or laterites, so called because they are formed by weathering or laterization), each of which requires a different technology for recovering the contained nickel from the ores. Known sulphide ores are located primarily in a few areas, notably Canada, the Union of Soviet Socialist Republics, Finland, Australia, Zimbabwe, the Republic of South Africa and Botswana, and often contain copper as well as nickel. Lateritic ores generally occur in tropical and subtropical regions, primarily in developing countries including New Caledonia, Indonesia, Cuba, the Philippines and Brazil and may contain cobalt or minor concentrations of other metals in addition to nickel.

Both sulphide and lateritic ores have been exploited for many years. While the sulphides exploited in earlier periods had high average grades of nickel, average grades currently being mined in such deposits are often lower than the nickel grade in laterite deposits. Sulphide ores are economically more profitable to work than laterites, however, because of their by-product content, their higher recovery rates and the lower costs associated with processing such ores. Ore grades and by-products in select major deposits of nickel (both sulphides and laterites) are summarized in Table 4.

About forty-five per cent of the reserves in producing mines contain sulphide ores but the share of lateritic resources in deposits that are not presently being worked but considered economically exploitable at the present time is over ninety per cent.

The most important use for nickel is in the steel industry, mainly in the production of stainless steels and other alloys. Its use in stainless steels accounts for 75% of the total consumption of nickel metal. Electroplating, chemical and other industrial uses account for the remaining 25%. World consumption of nickel in 1980 was 725,200 tons. Nickel consumption is closely related to steel production and, in general, to the capital goods sector and consumer durables. During the 1980s and 1990s, the evolution of the nickel market is expected to resemble that of steel, especially stainless steel and the iron and steel castings subsector. The slowdown of economic capacity during the 1970s considerably affected these sectors as reflected in the decrease of the growth of steel production. In conjunction with these developments, nickel consumption which had been growing at 6.7% per annum in the 1960s, decelerated to 2.6% per annum during the 1970s. During the decades of the 1980s and 1990s, nickel consumption is expected to grow in line with steel production. Even though the growth rate of nickel consumption by country/region is expected to be rather similar to that of steel, due to regional differences in nickel intensities, the expected overall growth between 1980 and 1995 is forecasted to lie between 2.5 and 3.0%.[5]

As known sulphide ores become depleted, lateritic nickel ores are expected to become more important. In the longer term (1995

Analysis of Processing Technology

to 2000), nickel prices have been projected as a function of costs of production for new projects. Lateritic ores, which are more expensive to process when compared to sulphide ores, seem to be the high marginal cost producer and therefore, the main determinant of future nickel prices.

Table 4. Ore Grades and By-products in Major Nickel Deposits

Country/Deposit(s)	Type	Nickel Grade (in percent)	By-product/Grade (in percent)
1. CANADA			
Inco-Thompson	Sulphide	2.0–2.2	Cobalt/0.15–0.2
Inco-Sudbury	Sulphide	1.41	Copper/0.97
2. AUSTRALIA			
Kalgoorlie-Kambalda	Sulphide	3.23	
Spargoville	Sulphide	2.47	Copper/0.23
Greenvale	Laterite	1.57	Cobalt/0.12
3. BOTSWANA			
Selebi	Sulphide	0.7–0.9	Copper/1.3–1.6
Pikwe	Sulphide	1.1–1.45	Copper/1.1
4. NEW CALEDONIA			
Thiébaghi	Laterite	3.0	Cobalt/
Poum	Laterite	2.3	—
Goro	Laterite	1.6	—
5. CUBA			
Moa Bay	Laterite	1.35	Cobalt/0.14
Nicaro	Laterite	1.4–1.8	Cobalt/0.1
6. PHILIPPINES			
Surigao	Laterite	1.2	Cobalt/0.1
7. INDONESIA			
Soruako	Laterite	1.76	Cobalt/0.08–0.15
Gag Island	Laterite	1.52	Cobalt/0.15
8. COLOMBIA			
Cerro Matoso	Laterite	1.5–2.6	
9. GUATEMALA			
Lake Izabel	Laterite	1.75–2.1	—
10. USSR			
Norilsk	Sulphide	0.5	Copper/0.8

Source: *The Nickel Industry and the developing countries.* UNDTCD ST/ESA/100.

COPPER

Copper is one of the most useful and versatile metals and is used both in pure form and in alloys such as brass (with zinc), bronze (with tin), and nickel-silver (with zinc and nickel). Good electrical and thermal conductivities, resistance to corrosion, ductility and malleability, high strength, lack of magnetism and a pleasing reddish colour are properties of copper that are the basis for its vast industrial applications.

Copper deposits tend to be concentrated geographically. Currently, world copper reserves are estimated at 505 million tons (Table 5). Seven countries with the largest reserves (Chile, USA, USSR, Zambia, Canada, Peru and Zaire) account for 70% of existing world reserves. The next four account for another 12%.

Table 5. World Copper Reserves

Countries/Economies	Reserve (million tons)
INDUSTRIAL & DEVELOPING	448
Australia	16
Canada	32
United States	90
Chile	97
Papua New Guinea	14
Peru	32
Philippines	18
South Africa	5
Zaire	30
Zambia	34
China	3[a]
Other	77
CENTRALLY PLANNED	57
Poland	13
USSR	36
Other	8[a]
WORLD	505

[a]Estimate.

Source: US Bureau of Mines, *Mineral Commodity Summaries 1982.*

World consumption of copper peaked in 1979 to 9.82 million tons but then declined in subsequent years mainly because of the stagnation in the world economy. Historically, world consumption of refined copper has grown at a long-term average rate of 4.3% per annum (1950–1980). In the long-term three broad factors are adversely affecting consumption of refined copper; (a) rates of economic growth; (b) substitution by other materials such as aluminium and plastics, and (c) material-saving innovations in the end-uses of copper.

Taking into account all of these factors, world consumption is expected to grow at an average growth rate of between 2.7% and 3.0% per annum.

COBALT

Cobalt has a number of specialized uses, in heat, corrosion-resistant and tool steels, in hard-facing material for drilling equipment, in the manufacture of permanent magnets and in the chemical industry. The considerable increase in the price of cobalt in 1979 led to the substitution of ceramics in permanent magnets and there are other areas where substitution can occur. However, with the return to a more stable price for cobalt it is possible that an increasing demand will develop.

The present world consumption of cobalt is in the order of 33,000 tons per year, of which Zaire supplies about 50%. The United States Bureau of Mines has estimated that the future annual demand for cobalt will increase at a rate of about 3.0%.

Cobalt occurs in several different mineral forms, occasionally on its own but usually in association with copper and nickel ores. The amount of cobalt in these minerals is small, varying from less than 0.1% up to more than 0.5% and the physical association with the other metals is very close, with the result that the separation process is complex and costly. Cobalt is considered as a co-product or a by-product and processing plants are primarily designed to obtain the highest possible recovery of the main metals, copper and nickel. However, in pursuing this objective, it is not always economically sound or even technically practicable to aim for an equally high recovery of cobalt. It would be very difficult to obtain reliable statistical information on over-all cobalt recovery but, as an estimate, it could be said that, of the cobalt-bearing ore presently being treated, at least 90% of the copper and the nickel is extracted but probably less than 50% of the cobalt.

The factors which control the amount of cobalt that is produced, in the short term, are the type of copper- and nickel-processing plant in use and the capacity of the cobalt recovery sector of the plant: it is not necessarily a direct function of the amount of copper or nickel produced. In the case of some cobalt-bearing copper and nickel ores, no processing plant has been installed to recover the cobalt and it is discarded, mainly in smelter slag.

The definition of cobalt as a by-product is not strictly correct. Certainly, a cobalt-bearing material emerges from the copper and nickel recovery plants but the further processing to a saleable product entails an additional highly capital intensive recovery plant. The decision that then faces the producers (land-based as well as sea-bed) is whether the market demand and price justify the use of a mineral processing route which allows for recovery of cobalt, which could entail a lower recovery of the other metals, and the extra capital cost of the cobalt recovery section. The costing structure of multi-product industries is complicated and it would not be possible in this study to try to detail a comparison of relative costs between different land-based plants and the likely cost of sea-bed production. However, it is conceivable that, by the time sea-bed production of cobalt affects the market, the land-based cobalt plants will be financially written off and will therefore have a cost advantage over sea-bed production, particularly in the early years while depreciation is still a charge.

Table 6 summarizes data on world production and reserves of cobalt. The countries listed represent the current producers of cobalt and own all of the present mine capacity.

METAL PRODUCTION FROM NODULES AND MARKETING STRATEGIES

As noted in Table 2, and in order of importance in terms of value, it would appear as if the preferred order for metal production from nodules should be manganese, nickel, copper and cobalt. Following the review of the markets for these metals however, the decision is not as clear cut.

In respect of manganese, it has been noted that its market is highly integrated with the steel industry and that the ore reserves and supplies of manganese ore are ample but geographically

Table 6. World Production and Reserves of Cobalt

From: Lévy, J. -P. and N. A. Odunton, 1984. Economic impact of seabed mineral resources development in light of the Convention on the Law of the Sea, *Nat. Res. Forum*, **8**, no. 2.

Country	Cobalt content of 1980 mine production (10³ lbs.)[a]	Mine production capacity 1985 (10³ lbs.)[b]	Estimated reserves, 1983 (10³ metric tons)[c]
Zaire	34180	42000	2086
Zambia	9700	12000	544
USSR	4740	5000	227
Cuba	3580	4000	907
Canada	3534	5000	259
Australia	3520	4000	91
Philippines	2934	3000	399
Finland	2282	3500	34
Morocco	1848	2500	5
Botswana	498	1000	27
New Caledonia	400	2000	862
USA	—	—	862
South Africa	200	500	68
Indonesia	—	—	544
Greece	—	—	125
(A) Total	67416	84000	7040
(B) World total	75476	84000	7312
(A) as per cent of (B)	89.3	100	96.3

Sources: [a] USBM, 1981.
        [b] USBM, 1980.
        [c] Kirk, 1983.

concentrated in the Union of Soviet Socialist Republics and the Republic of South Africa. Entry into the manganese market will, therefore, be difficult. Furthermore, because the grades of nodules from the Pacific are higher than those from the Atlantic, operations to extract valuable metals from nodules will in all probability be concentrated on the Pacific coast and, therefore, will be badly located with respect to the largest US steel producers. Manganese consumption in the long term is also expected to expand at an average rate of 2.7% per annum.

The value of nickel contained in the nodules is next after manganese. The nickel corresponds to over twice the sum of copper plus cobalt on a revenue basis. Moreover cost effective substitutes for the major nickel applications are not available. Finally, the primary new source of nickel other than deep sea polymetallic nodules are the laterites. As previously mentioned, lateritic ores are more expensive to process meaning that production costs for extracting nickel metal from them are higher than the cost of processing sulphides. If nickel from nodules can be extracted at a price competitive to that from laterites, then nickel as a prime recovery target would have much to recommend it.

The degree of market penetration required of a typical nodule mine is a further important consideration. The potential output of a 10,000 ton per day nodule mine is shown in Table 7.

Table 7. Comparison of 10,000 ton per day (3,000,000 ton per year) nodule mine output with estimated 1985 metal markets.

Metal	Annual Mining Rate tons	Percent of World Production
Manganese	810000	4.5%
Nickel	39000	4%
Copper	33000	0.3%
Cobalt	6000	46%

Source: Lévy, J-P., and Odunton, N.A., Economic impact of sea-bed mineral resources in light of the Convention on the Law of the Sea. *Natural Resources Forum*, **8**, no. 2, 1984.

An analysis of the table indicates the following:

(i) Any commercial nodule venture locked into the production of manganese to make a profit would be risky. The primary reason is that a single operation will supply a major portion (in this case even more) of the anticipated average annual growth rate. As a source of manganese, nodules will only become important as terrestrial sources of manganese become depleted;

(ii) Market penetration in the case of nickel is not as pronounced. Moreover, if manganese is considered unattractive at the present time, then nickel would be responsible for about seventy percent of the revenues from manganese nodule mining. Against this, however, the caveat must be thrown in that worldwide there are ample reserves of nickel ore.

METAL PRODUCTION AND STRATEGIC CONCERNS

The arguments provided above indicate that in the long term, purely economic concerns may become the catalyst for nodule mining. At the present time however the arguments for nodule mining appear to stem more forcefully from perceived strategic concerns.

From the literature on the economics of nodule mining (including transport to land), the cost item of prime concern is the mining cost because the downstream charges for processing are about the same for all metal producers whether on land or on the sea. The major cost difference among producers is the mining cost; this, however, is only a small portion of the total cost of metal. For most producers, including the potential nodule miners, the mining cost ranges between 10 and 25 percent of the total operating cost. Most of the cost of producing metal from nodules is hydrometallurgical processing, which amounts to between 40 and 67 percent of the total operating cost. A percentage reduction in the operating cost would have the greatest impact if hydrometallurgical processing techniques could be improved. However, metallurgical improvements would also be applicable to minerals derived from land based sources, thereby preserving their current advantage.

The great capital cost of the project is the chief cause of low profitability or losses. Can capital cost be significantly reduced? This question cannot be answered at the current stage of technological knowledge. The system for mining has not been designed, and, after design and construction, it will still need testing and experience in use before it becomes completely acceptable. The hydrometallurgical system is at the same state of development.

In the event of capital losses, and in some cases operating cost losses to acquire manganese, copper, nickel, or cobalt, the primary issue is the willingness of governments to accept such losses. Since a nodule mine could satisfy the requirements of many countries for nickel, cobalt and manganese, political motivations, considerations of national security and internal economic stability may overweigh economic concern over capital losses. A country may indeed construct a nodule mining and processing system merely to ensure that it will have a source and control of metal supply.

From the previous discussion, it should be clear that there is no such thing as an "optimum" processing route for manganese nodules. Whatever processing alternative is finally chosen by a consortium, it will be a compromise. This means that it will be deficient in comparison to some other process in one or more respects; however, in terms of the major objectives of the consortium and its overall economic and financial benefits, it will be the preferred alternative. Also, since the decision to commit resources to the construction of a plant is some years away, the preferred alternative could change as metal prices or underlying factors change.

NOTES

1. (a) Wills, B.A., 1981. *Mineral processing Technology*. Second edition, Pergamon, Oxford, 525 pp.
   (b) Kelly, E.G. and D.J. Spottiswood, 1982. *Introduction to mineral processing*. Wiley-Interscience, New York, 491 pp.
2. (a) Pehlke, R.D., 1973. *Unit processes in extractive metallurgy*. Elsevier, New York.
   (b) Gilchrist, J.D., 1980. *Extraction metallurgy*. Second edition, Pergamon, Oxford.
3. (a) *Gravity concentration* is a method of separating solids of different specific gravities in a fluid medium—often water or air.
   (b) *Flotation* is a complex physicochemical process taking place in an ore pulped with water, by which the surfaces of one or more minerals in the finely ground pulp are made water-repellent and responsive to attachment with air bubbles. Separation of prepared pulps takes place in flotation machines, which are essentially open troughs with a means of introducing and dispersing air into the pulp. The mineralized air bubbles or froth collected in the rougher separation customarily are cleaned by routing through additional flotation machines to drop the misplaced, poorly liberated or poorly reagentized minerals. These middlings are routed within the circuit to reclaim additional values. The cleaned products are *concentrates* and the waste products are contained in the *tailings*.
   (c) *Hydrometallurgy*, a subdivision of extractive metallurgy, includes a diversity of processes featuring the selective dissolution of metals from ores and concentrates and the subsequent recovery of relatively pure metal compounds or metals.
   (d) *Pyrometallurgy*, also a subdivision of extractive metallurgy, encompasses all processing operations using refractory-lined furnaces and high temperatures created by electrical energy or burning fuels to produce refined metals from ores and concentrates.
4. Some countries, such as the USSR, use above average levels of manganese in their steel production. The main reason is ample domestic supplies of manganese ore. Other countries which depend on manganese imports use it more sparingly. The quantity of manganese used in the production of steel depends

not only on the particular specification for the steel but also on the relative prices of iron ore and manganese.

5. United Nations Department of Technical Co-operation for Development, *The Nickel Industry and the Developing Countries*. (N.Y. 1980) ST/ESA/100.

Chapter 3

Unit Operations of Extractive Metallurgy and their Applicability to Nodule Processing

The extraction of metals is a manufacturing process whose raw material is typically an ore and whose output is a refined metal. The various steps or unit operations, that form the tie linking the raw ore to the refined metal are characterized by their ability to effect various specialized kinds of separations. In the initial stages, the separations tend to involve primarily physical techniques and the various operations fall under the discipline of mineral processing. On the other hand, the later stages are largely dependent on the application of chemical techniques and the corresponding processes constitute the discipline of chemical or extractive metallurgy.

MINERAL PROCESSING OPERATIONS

The first mineral processing unit operation involves the application of comminution (i.e. crushing and grinding) techniques. The run-of-mine ore is usually a mixture of minerals, and therefore, provided the metal values occur as discrete minerals, this operation serves to liberate the valuable minerals from the unwanted (gangue) minerals. Where the metal values do not occur as discrete minerals, the comminution operation may still be needed to produce fine particles which are more amenable to the subsequent processing operations. The primary liberation of ores is achieved during mining. In the next stage of size reduction, the run-of-mine ore is subjected to crushing which involves the compression of individual rocks ranging in size from several feet to a few inches. Subsequent size reduction of the coarsely crushed product by impact, shearing or inter-particle attrition is achieved by grinding. Further reduction in size to fine particles is achieved by pulverizing which also involves impact, attrition and shear. Comminution can be achieved either with dry solids (crushing) or when the material is in a slurry with water (grinding and pulverizing).

Following comminution are the concentration operations which take advantage of differences in the physical and surface chemical properties of the valuable and gangue minerals to produce an enriched fraction (concentrate) containing the valuable metal, and a waste fraction containing the gangue minerals. Common concentration operations include sieving, magnetic separation and flotation.[1] The most important commercial concentration technique is froth flotation. This technique is based on the ability to modify the affinity of fine particles for air bubbles in aqueous media, by the introduction of soap-like compounds to the aqueous phase. The ability to produce a concentrate early in the metal extraction process cuts down on the size of subsequent plant stages needed for chemical processing and, therefore, leads to significant savings.

CHEMICAL METALLURGICAL OPERATIONS

Once a concentrate has been prepared, the next stage in the separation process involves the application of chemical methods to break the metal–nonmetal bonds present in the ore minerals This aspect of the extraction process relies on various kinds of hydrometallurgical (e.g. leaching, solvent extraction, electrowinning) and pyrometallurgical (e.g. roasting, smelting, converting techniques.

Figure 1 presents a simplified flowsheet (i.e. a conceptual diagram summarizing the sequence of operations) for a hydrometallurgical metal extraction process. Figure 2 is the corresponding diagram for a pyrometallurgical process.

HYDROMETALLURGICAL OPERATIONS

Hydrometallurgy includes a diversity of processes featuring the selective dissolution of metals from ores and concentrates and the subsequent recovery of relatively pure metal compounds or metals. The unit operations comprising hydrometallurgical pro

Fig. 1. Unit Operations in Hydrometallurgical Processing

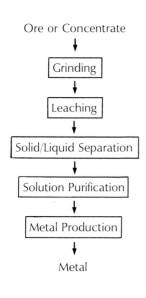

Fig. 2. Unit Operation in Pyrometallurgical Processing and Electrorefining

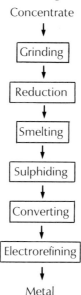

cesses invariably include preparation of the feed material (grinding etc.), leaching with an aqueous solution of an acidic or alkaline solvent, separation of the metal-bearing solution from the leach residue, and purification of the solution followed by the recovery of the purified metal compound.

Hydrometallurgical processing of ores and concentrates, except for the rare application of in-place leaching, requires preparation of the raw materials to assure effective leaching and product recovery. Ores must be crushed — and usually ground and sized — prior to leaching to permit effective contact between the ore minerals and the solvent. Although the oxide and carbonate minerals of many of the important metals are readily dissolved, sulphide and silicate minerals must be treated to convert them to soluble form. Sulphide minerals may be roasted using air and/or chlorides to produce soluble oxides, sulphates or chlorides. Silicate and insoluble oxide minerals are roasted with reductants, chlorides, sulphur compounds or soda ash.

The dissolution of minerals in any of a variety of lixiviants is called leaching. Dilute, low-cost sulphuric acid is the most important solvent used for leaching ores and concentrates. The chief alkaline solvents are the hydroxides and carbonates of sodium and ammonium. Ammonium carbonate solution is used for leaching certain copper and nickel minerals. The most important of the various leaching methods is agitation of finely ground material in open vessels at atmospheric pressure. Vat, heap and dump leaching comprise the other methods used to extract metals from ores.

In general, the degree of extraction that can be attained is dependent on the physical and chemical nature of the raw material, the length and temperature of the leach, and the type and strength of the solvent used. In order to minimize reagent consumption, it is often desirable that the dissolution be selective (i.e. attacking only the valuable metals). The undissolved material is separated from the metal-containing (or pregnant) solution by dewatering operations such as sedimentation and filtration. Because no solvent is completely selective, pregnant solutions contain impurities that must be removed to make a metal product of acceptable quality. Therefore, the next step, solution purification, serves to:

(a) purify the pregnant solution by removing undesirable metals;

(b) separate one valuable metal from another; and,

(c) upgrade the metal content in solution (i.e. increase the metal concentration) in preparation for the subsequent (downstream) operation of metal production.

Solution purification and product recovery usually require a combination of techniques. These may include evaporation or cooling to promote selective crystallization, use of chemicals to selectively precipitate impurities, application of ion exchange and solvent extraction, or reduction of dissolved metal salts to metal powder using hydrogen or other reductants.

The most significant advance in solution purification technology in the last twenty years has been the commercial application of solvent extraction techniques to the recovery of copper and also of nickel and cobalt.[2] In selective solvent extraction the metal-containing aqueous solution is contacted with a water insoluble organic solvent (kerosene) containing an organic compound (extractant) capable of reacting with a particular aqueous phase metallic species. After allowing the organic and aqueous phases to separate, the metal extracted into the organic phase is extracted back, (i.e. stripped) by contacting the metal-loaded organic phase with an aqueous solution whose composition forces a reversal of the previous extraction reaction. By appropriately adjusting the loaded organic/strip solution volumetric ratio, it is possible to produce a strip product liquor whose metal concentration is several times higher than that in the initial pregnant solution.

Once a purified solution has been obtained, a metallic product can be precipitated by using a variety of techniques such as electrolysis (electrowinning) and hydrogen reduction. In some cases, a market product of a metal salt is obtained by crystallizing the appropriate salt from the purified solution.

PYROMETALLURGICAL OPERATIONS

Pyrometallurgical processing encompasses all operations using refractory-lined furnaces and high temperatures created by electrical energy or burning fuels to produce refined metals from ores and concentrates. Drying, roasting, sintering, distilling, smelting and fire refining techniques are the major unit processes employed. Drying, roasting and sintering are subsidiary unit operations that are widely employed to prepare ores and concentrates for smelting.[3] Smelting refers specifically to those high-temperature processes whereby ore and gangue minerals are chemically altered, fluxed and reduced to form a low-density molten slag and one or more heavier liquid metals or metallic compounds.

For example, in the pyrometallurgical processing of sulphide concentrates, the roasting step (500–700°C) causes partial oxidation of the sulphides to oxides and sulphates and partial removal of sulphur as SO_2.[4] The roast product, calcine, is smelted (liquefied) at about 1200°C with the addition of silica, to give a liquid sulphide phase (matte) which contains the copper, sulphur and unoxidized iron. The waste material which comprises mostly the oxides of silicon, aluminium and iron, forms a molten oxide phase (slag). Since the molten matter is heavier than, and practically immiscible with, the slag, the two liquids are readily separated. Smelting is accomplished in a variety of equipment, including reverberatory, blast, electric and flash furnaces.[5] Converting follows matte smelting. In this operation, the molten matte is treated with air in the presence of silica to remove sulphur as sulphur dioxide (SO_2) and iron as iron oxide (which is then removed with the slag). The equipment usually used is a Pierre-Smith convertor, which is a horizontal refractory-lined cylindrical reactor. A crude metallic product (blister copper) analyzing at about 99% copper (Cu) is produced which is purified electrochemically (electrorefining) to give a product analyzing 99.99% + copper (Cu). Electrorefining is actually a hydrometallurgical process since it involves the aqueous dissolution (with the aid of electrical current) of the blister copper followed by the reprecipitation of pure copper onto copper cathodes.

All pyrometallurgical operations produce large volumes of gas containing a wide variety of vaporized metals, dust and fumes. Such processes often generate numerous by-product slags, drosses and metal spills that usually are recycled for reprocessing or sold to other reduction plants for recovery. Major facets of pyrometallurgy include the efficient utilization of heat energy; the design of many different types of high temperature furnaces, roasters, and kettles; the production of fluid discardable slags; and the selection of refractory furnace linings resistant to corrosion and erosion.

THE APPLICABILITY OF EXTRACTIVE METALLURGY TO NODULE PROCESSING

As noted above, the technology of metal extraction involves a succession of separation processes. In order for a commercially useful separation to be achieved, there must be significant differences between the physical and/or chemical properties of the gangue materials and those of the minerals containing the metal values. Ideally, before metal extraction procedures are proposed for an ore, detailed information must be collected on its chemical composition, mineralogy and texture (i.e. size and distribution of the valuable minerals). In practice however, it is often necessary to embark upon the task of process development without access to such detailed mineralogical and physical-chemical data. The process technologies for most land-based ores were discovered and implemented at a time when mineralogical analytical techniques were not well developed. In contrast, the commercial interest in polymetallic nodules has occurred at a period when modern analytical techniques such as scanning and transmission electron microscopy are being increasingly applied in mineralogical investigations. Nevertheless, as a result of the complexity of manganese nodules, the chemical and mineralogical characteristics are as yet not fully elucidated.[6]

NOTES

1. Magnetic separation comprises sorting one solid from another by means of a magnetic field and is based upon the principle that particles placed in a magnetic field are either attracted (paramagnetic) or repelled (diamagnetic) by it. Flotation is a complex physicochemical process taking place in an ore pulped with water, by which the surfaces of one or more minerals in the finely ground pulp are made water-repellent and responsive to attachment with air bubbles.
2. Flett, D. S., 1982. Solvent extraction in hydrometallurgy. In K. Osseo-Asare and J.D. Miller, eds., *Hydrometallurgy. Research, Development and Plant Practice.* T.M.S.–A.I.M.E., Warrendale, Pa. pp. 39–64.
3. (a) Drying is a low-temperature operation done in furnaces designed to remove excess water efficiently.
   (b) Roasting employs heat in hearth, fluid-bed, flash and rotary-kiln-type furnaces to alter, without fusion, the chemical form of minerals in the raw materials in oxidizing, reducing or neutral atmospheres.
   (c) Sintering and closely related nodulizing differ from roasting in that higher temperatures are used to effect agglomeration of fine material particles by incipient fusion as well as by chemical change.
4. Biswas, H.K. and W.G. Davenport, eds. *Extractive Metallurgy of Copper.* Second edition, Pergamon , Oxford.
5. *ibid.*
6. Glasby, G.P., ed., 1977. *Marine Manganese Deposits.* Elsevier, Amsterdam.

Chapter 4

Physical, Chemical and Mineralogical Characteristics of Polymetallic Nodules

PHYSICAL CHARACTERISTICS

Polymetallic nodules are characterized by a very high porosity of about 60% which gives this material an apparent density of 1.4 g/cm³ compared with a real density of 3.5 g/cm³ on a pore-free basis.[1] A study of several different nodule samples by Fuerstenau *et al.* revealed that over 80% of the total pore volume is occupied by pores whose diameter lies in the range of 0.01 to 1.0 microns (10^{-4} cm.)[2]. Thermal treatment of nodules up to 850°C has little effect on the total porosity; however, there is a systematic increase in pore size (i.e. the finer pores collapse, leaving wider pores). Thus, whereas a nodule sample dried at 110°C gave a pore size distribution of 85% finer than 0.1 micron, thermal treatment at 850°C resulted in pores which were only 5% finer than 0.1 micron. It is likely that such changes in internal structure would affect the rates of roasting processes (e.g. the reductive roast processes utilizing hydrogen (H_2), carbon monoxide (CO) and hydrogen chloride (HCl)). The fine internal pore structure of nodules also accounts for their very high specific surface areas with typical values of 200 m²/g.[3] Given their high porosity and their occurrence in the marine environment, it is not surprising that nodules carry a high water content. Fuerstenau *et al.* have reported water contents in the range of 18.6% to 25.7% on a wet basis.[4]

The presence of a high amount of water in nodules would be expected to have an adverse effect on the efficiency and cost of processing schemes that require prior drying of nodules. As indicated in Table 8, when nodules are heated in the temperature range of 100°–1000°C, between 40% and 70% of the total heat supplied is used to evaporate water.[5]

Table 8. Energy Requirements for Heating One Mole (41 grams) of Nodules from Room Temperature to Various Temperatures

	TEMPERATURE (°C)					
	100	200	400	600	800	1000
1. Total heat required (Cal.)	1330	2540	4960	7490	9920	12550
2. Heat required for water only (Cal.)	920	1500	2480	3460	4440	5410
3. Percentage of total heat requirements for water evaporation	69%	59%	50%	46%	45%	43%

From: Han, K.N., and D.W. Fuerstenau, 1975, Acid leaching of Ocean manganese nodules at elevated temperatures. *International Journal of Mineral Processing*, **2**, pp. 163–171.

An important consideration in any metal extraction process is the energy consumption. Comminution is generally a costly unit operation; for example, in the production of 1 tonne of metallic copper from feed ore material analyzing 0.6% copper, almost a third of the total energy required for the process is utilized in the comminution operation (33×10^3 kWh/tonne of metal).[6] Fortunately, nodules are friable and easy to grind. Brooke and Posser have reported a relatively low grindability index (i.e. the energy needed to grind 80% of a nodule sample to a size smaller than 100 mesh) of about 10 kWh/ton.[7]

CHEMICAL AND MINERALOGICAL CHARACTERISTICS

Manganese nodules are oxide ores, the main mineralogical phases being a variety of manganese and iron oxides.[8] The major manganese minerals are:

(i) todorokite (an oxide of manganese (Mn), magnesium (Mg), calcium (Ca), sodium (Na), and potassium (K) which can be described by the chemical formula (Ca, Na, K) (Mg, Mn^{2+}) (Mn_5^{4+} O_{12} . $3H_2O$);

(ii) buserite (or 10 Å manganite, a sodium manganese oxide hydrate),

(iii) birnessite (or 7 Å manganite) and,

(iv) vernadite (or $-MnO_2$).

The crystal structures of these manganese minerals have not fully been established. However, these structures may be approximated by thinking of them in terms of layers of $[Mn^{2+}(OH, H_2O)_2]$ sandwiched between sheets of edge-shared $[Mn^{4+}O_6]$ octahedra. As a result of manganese deficiencies in the $[Mn^{4+}O_6]$ layers, Mn^{2+} and Mn^{3+} ions may substitute for Mn^{4+}.

The iron oxides that have been identified include goethite ($-FeOOH$) and hematite (Fe_2O_3). Like the manganese oxides, the fundamental structural unit of the iron oxides in nodules is the $[Fe^{3+}O_6]$ octahedron. The manganese and iron oxides occur in a very fine state of subdivision. For example, Von Heimendahl *et al.* have identified a ferric oxide phase in the 30–60Å size range.[9]

No discrete minerals of copper (Cu), nickel (Ni), or cobalt (Co) have been identified in nodules. It is likely that these trace elements occur as substitutes for Mn^{4+}, Mn^{3+}, Mn^{2+} and Fe^{3+} in the various manganese and iron oxide minerals.[10] Other possibilities include their occurrence as adsorbed species on the manganese and iron oxide surfaces or as intimate colloidal size admixed co-precipitates with the Mn and Fe oxides. There is some evidence for the possible enrichment of specific trace metals in the Fe and Mn oxide phases. For example, in a selected nodule sample examined by Burns and Fuerstenau, two manganese oxide regions containing MnO_2 and 7Å manganese were identified over the sample cross-section. The 7Å manganite region was rich in nickel and copper, whereas the MnO_2 region showed an enrichment in iron and cobalt.[11] Nevertheless, the available evidence indicates that the spatial distribution of metals in nodules as well as the inter-element association varies from sample to sample and no generalizations can be made at this stage in our knowledge of their mineralogy.

PHYSICAL AND CHEMICAL CONSTRAINTS IN METAL EXTRACTION SCHEMES FOR NODULES

A major consideration in the development of suitable metal extraction procedures is the fact that no discrete copper, nickel or cobalt minerals have been identified in nodules. Like most low grade oxidized ores, therefore, physical upgrading is difficult if not impossible. It was noted above that some investigators have reported the enrichment of certain trace elements in the iron and manganese oxide phases. Even if this were a widespread phenomenon (which appears to be doubtful), the fact that the manganese and iron oxides occur as extremely minute particles in the 100Å size range precludes any useful application of such physical separation and concentration techniques as grinding (to achieve liberation) and flotation.

Since the production of a nodule concentrate is out of the question, the only alternative left is direct chemical processing. If, as current evidence seems to indicate, the trace elements are located within the crystal structures of the manganese and iron oxide minerals, then it is necessary to destroy these host mineral structures in order to release the contained copper, nickel and cobalt. Thus, the extractive metallurgy of manganese nodules is based on the exploitation of chemical properties of both the contained valuable trace metals such as copper, nickel and cobalt and the manganese-iron oxide matrix minerals. Both pyrometallurgical and hydrometallurgical extraction methods have been proposed to effect the separation.

NOTES

1. The *porosity* of a rock is defined as the percentage of total volume occupied by pore spaces.
2. (a) Fuerstenau, D.W. and K.N. Han, 1983. Metallurgy and processing of marine manganese nodules, *Mineral Processing Technology Review*, 1, pp. 1–83.
   (b) Fuerstenau, D.W. and K.N. Han, 1977. Extractive Metallurgy, in G.P. Glasby, ed., *Marine Manganese Deposits*, Elsevier, Amsterdam, pp. 357–390.
3. (a) *ibid.*
   (b) Fuerstenau, D.W., A.P. Hefring, and M. Hoover, 1973. Characterization and extraction of metals from sea floor manganese nodules, *Transactions of the Society of Mining Engineers (SME)*, 254, pp. 205–211.
4. Fuerstenau, D.W. and K.N. Han, *op. cit.*
5. Han, K.N., and D.W. Fuerstenau, 1975. Acid leaching of ocean manganese nodules at elevated temperatures. *International Journal of Mineral Processing*, 2, pp. 163–171.
6. (a) Wills, B.A., 1981, *Mineral Processing Technology*, Second edition, Pergamon, Oxford, 525 pp.
   (b) Kelly, E. G. and D.J. Spottiswood, 1982. *Introduction to Mineral Processing*, Wiley-Interscience, New York, 491 pp.
7. Brooke, J. N. and A.P. Prosser, 1969. Manganese Nodules as a Source of Copper and Nickel—Mineralogical Assessment and Extraction. *Trans. IMM*, 78C, pp. 64–73.

8. Burns, R. G. and V.M. Burns, 1977. Mineralogy, in G.P. Glasby, ed., *Marine Manganese Deposits*, Elsevier, Amsterdam, pp. 185–248.

9. von Heimendahl, M., G.L. Hubred, D.W. Fuerstenau and G. Thomas, 1976. A Transmission Electron Microscope Study of Deep-Sea Manganese Nodules, *Deep-Sea Res.*, **23**, pp. 69–79.

10. Burns, R. G. and V.M. Burns, 1977. Mineralogy, in G.P. Glasby, ed., *Marine Manganese Deposits*, Elsevier, Amsterdam, pp. 185–248.

11. Burns, R.G. and D.W. Fuerstenau, 1966. Electron-probe Determination of Inter-element Relationships in Manganese Nodules, *Amer. Miner.*, **51**, pp. 895–902.

Chapter 5

Metal Extraction Schemes

PYROMETALLURGICAL PROCESSING

As a result of the complexity and polymetallic nature of nodules, it is not possible to use purely pyrometallurgical techniques to produce separate metallic products of the contained metal values. However, as a pretreatment step, pyrometallurgy can be used to:

(i) concentrate the metal values prior to hydrometallurgical processing,
(ii) alter the physical, chemical and mineralogical characteristics to achieve enhanced selectiveness and dissolution rates and
(iii) decrease the total amount of material sent to leaching.

Figure 3 and Table 9 present a summary of the various pyrometallurgical approaches that have been investigated.[1] In Scheme 1, there is no high temperature pretreatment. Ground nodules are subjected to leaching as the first step in the process. In Scheme 2, dried, ground nodules are subjected to roasting which is conducted under reducing conditions at temperatures in the neighbourhood of 600°C. A variety of reductants such as coke (carbon), gaseous hydrogen and carbon monoxide may be used.[2] A finely divided metallic product that is intimately admixed with reduced oxides of manganese and iron is obtained. Since the physical separation of the metallic and oxide phases of the product is not possible, the entire roast product is subjected to the subsequent leaching stage. In Scheme 3, the reductive roasting step is conducted at temperatures considerably higher (1000°C) than in Scheme 2 and the roast product is given a further pyrometallurgical treatment, that is smelting.[3] The smelting step is a melting operation which permits two products to be obtained: a metallic

Fig. 3. Pyrometallurgical Pretreatment Methods

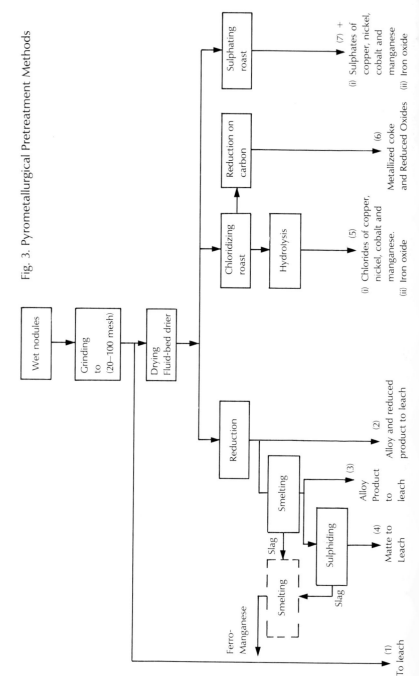

Table 9. Pyrometallurgical Processes

Nature of High Temperature Pretreatment (Schemes)	Process Conditions	Characteristics of Product	Destination of Product
(1) None			Directly to leaching
(2) Reductive roasting	Reduction temperature range — 350–800°C. Feed size sent to reduction between −100 to −20 mesh Reductant gas including carbon monoxide (CO), gaseous hydrogen (H_2) and carbon (C).	Metallic grains of Cu, Ni, Co, finely disseminated in a matrix of the reduced oxides of manganese and iron (MnO, FeO); Partial or complete alloying with some metallic iron; Physical separation into Cu, Ni, CO concentrate not possible; Good solubilization of copper (80%–90%), nickel (80–90%), cobalt (50–70%).	Alloy and reduced product goes to leaching
(3) Smelting to produce alloy[a]	Reductive roast temp. (900–1000°C). Smelting temperature greater than 1400°C. Time for operation (0.3–2 hrs.) Carbonaceous reductants (graphite, coke and mixtures of carbon monoxide, hydrogen, carbon dioxide and water).	Selective recovery of copper, nickel and cobalt in the metallic product. Recovery in the alloy phase is 85–95% for copper, 93–98% for nickel, 90–98% for cobalt, 80–90% for iron, and 0.5–2.5% for manganese. Slag phase containing most of the manganese and some iron.	Alloy phase goes to leaching. Manganese in the slag phase can be sent to smelter for conversion to ferromanganese.

[a] Vasil'chikov, N.V., G.B. Shiner, I.F. Krasnykh and E.A. Grishankova, 1968. Iron-manganese nodules from the ocean floor raw materials for the production of cobalt, nickel, manganese and copper. *Tsvet. Metal,* 41 (1), pp. 40–2; Beck, R.R. and M.E. Messner, 1970. Copper, nickel, cobalt and molybdenum recovery from deep sea nodules, in R.P. Ehrlich, ed., *Recovery Metallurgy,* AIME, New York; Sridhar, R., 1974. Thermal upgrading of sea nodules. *Journal of Metals,* December 1974, pp. 18–22; Sridhar, R., W.E. Jones and J.S. Warner, 1976. Extraction of copper, nickel and cobalt from sea nodules, *Journal of Metals,* April, 1976, pp. 32–37.

Table 9. (continued)

Nature of High Temperature Pretreatment (Schemes)	Process Conditions	Characteristics of Product	Destination of Product
(4) Smelting to produce matte	Molten alloy prepared as in Scheme 3 and reacted with air in the presence of silica to remove manganese. Purified alloy then reacted with alloy to produce matte. Iron in matte removed by reaction with air in the presence of silica (converting). Temperature range 1000–1420°C.	Copper, nickel and cobalt concentrated in matte product (approx. 5% dry nodule weight). Matte composition approx. 25% copper, 40% nickel, 5% cobalt, 5% iron, 20–25% sulphur. Slag phase containing most of the iron.	Matte product to leaching Slag phase to smelting for production of ferromanganese.
(5) Chloridizing roast[b]	Reaction of nodules with hydrogen chloride to give chlorides at temperature range of 500–800°C. Reaction of chlorides in temperature range 200–400°C with water to form water-insoluble iron oxide.	Particulate mixture of copper, nickel, cobalt and manganese chlorides and iron oxide. Further processing (i.e. leaching) uses entire roast product. By-product chlorine recoveries from water leach of roast product. 99.7% copper, 96.7% nickel, 97.0% cobalt, 99.9% manganese. Iron insoluble.	To leach
(6) Segregation roast[c]	Reaction of nodules with a chloride salt (e.g. NaCl, MgCl₂, CaCl₂) at about 1000°C in the presence of coke.	Oxides of manganese and iron. Metallic grains of copper, nickel, cobalt and iron alloy. May be concentrated by magnetic separation. Metal recover 46–88% copper, 23–32% nickel, 7–22% cobalt, 0–9% manganese, 0–9% iron.	To leach

[b] Kane, W.S. and P.H. Cardwell, 1973. Recovery of metal values from ocean floor nodules, U.S. Patent 3,752,745, 14 August; Cardwell, P.H., 1976. Method for separating metal constituents from ocean floor nodules, *US Patent 3,950,486*, 13 April.

[c] _____, 1975. Segregation roasting of nickel, copper and cobalt from deep-sea manganese nodules, *International*

Table 9. (continued)

Nature of High Temperature Pretreatment (Schemes)	Process Conditions	Characteristics of Product	Destination of Product
(7) Sulphating roast[d]	Reaction with 10% SO_2 and 90% air in temperatures above 650°C.	Particulate mixture of copper, nickel, cobalt and manganese sulphates and iron oxide. Further processing (i.e. leaching) uses entire roast product. Water leach of product yields 99% manganese, 87% copper, 92% nickel, 97% cobalt.	To water leach
(8) Sulphuric acid bake	Concentrated H_2SO_4 mixed with ground nodules. Mixture baked in temperature range 650°–700°C	- do -	To water leach

[d] Iwasaki, I., A.S. Malicsi and N.C. Jagolino, Segregation process for copper and nickel ores, *Progress in Extractive Metallurgy*, Vol. 1; Meyer-Galow, E., K.H. Schwarz and U. Boin, Metal extraction from manganese nodules by sulphating treatment, in *Interocean 1973*, Vol. 1, pp. 458–468; Van Hecke, M.C. and R.W. Bartlett. 1973. Kinetics of sulphation of Atlantic ocean manganese nodules, *Metall. Trans.*, 4, pp. 941-947. Kane, W.S. and P.H. Cardwell, 1975. Reduction method for separating metal values from ocean floor nodule ore, *US Patent 3,869,360*, 4 March; Baba, R. and T. Tamagawa, 1979. *Nippon Kogyo Kaishi*, 95, (1096), 353.

phase which contains most of the copper, nickel, cobalt and iron originally in the nodules and a slag phase to which most of the manganese and iron report. Scheme 4 takes the previous operation one step further. It uses the high temperature reaction of sulphur with cobalt, copper and nickel to separate these metals from iron.[4] The sulphiding step yields a matte product that contains the sulphides of copper, nickel and cobalt and a slag product containing most of the iron originally in the alloy product.

Scheme 5 takes advantage of the fact that copper, nickel, cobalt, iron and manganese readily form chloride salts characterized by low melting points, high volatility and high aqueous solubility. In addition, the chloride formed when iron is in the plus three valence state (i.e. ferric chloride, $FeCl_3$) readily decomposes in the presence of water-steam to give water-insoluble iron oxide (Fe_2O_3). This permits the selective extraction of copper, nickel, cobalt and manganese from iron by water leaching.[5] In the reductive chloridizing roast step, the application of hydrogen chloride gas to the nodule material at a temperature of 500–600°C results in the formation of chloride salts:

$$MnO_2 + 4HCl \longrightarrow MnCl_2 + 2H_2O + Cl_2 \tag{1}$$

$$FeOOH + 3HCl \longrightarrow FeCl_3 + 2H_2O \tag{2}$$

In the reaction described by equation (1) above, manganese in the plus four oxidation state (Mn^{4+}) is reduced to the plus two state (Mn^{2+}), with a corresponding generation of chlorine. On the other hand, the oxidation state of iron remains unchanged (Fe^{3+}). The hydrolysis step, conducted in a temperature range of 300–400°C, causes the conversion of ferric chloride to water-insoluble iron oxide and regenerates hydrogen chloride:

$$2FeCl_3 + 3H_2O \longrightarrow Fe_2O_3 + 6HCl \tag{3}$$

The high volatility of the chlorides of copper, nickel, cobalt, manganese and iron is used to advantage in Scheme 6, which is the segregation roast process.[6] In this method, a chloride salt (e.g. $NaCl$, $MgCl_2$, $CaCl_2$) is heated at about 1000°C with nodules in the presence of coke particles. Hydrogen chloride is generated *in situ* and the volatile metal chlorides are reduced at the coke surface to give metallic deposits.

Metal values in nodules can also be converted to sulphate salts

(Scheme 7) by reaction with gaseous SO_2 or by baking with concentrated sulphuric acid.[7] The reaction of manganese dioxide and goethite with gaseous sulphur dioxide can be represented by equations (4) and (5) respectively:

$$MnO_2 + SO_2 \longrightarrow MnSO_4 \qquad (4)$$

$$4FeOOH + 6SO_2 + 3O_2 \longrightarrow 2Fe_2(SO_4)_3 + 2H_2O \qquad (5)$$

Under the reaction conditions (i.e. temperatures greater than 650°C), ferric sulphate is unstable and subsequently decomposes:

$$Fe_2(SO_4)_3 \longrightarrow Fe_2O_3 + SO_3 \qquad (6)$$

Reaction (6) catalyzes the sulphation reactions since the resulting sulphur trioxide is also a sulphating agent:

$$CuO + SO_3 \longrightarrow CuSO_4 \qquad (7)$$

In the case of the sulphuric acid bake process (Scheme 8), there is an initial step (the pugging step) during which concentrated sulphuric acid is thoroughly mixed with ground nodules. The sulphation of the matrix oxides commences during this stage.

$$2FeOOH + 3H_2SO_4 \longrightarrow Fe_2(SO_4)_3 + 2H_2O \qquad (8)$$

$$2MnO_2 + 2H_2SO_4 \longrightarrow 2MnSO_4 + 2H_2 + O_2 \qquad (9)$$

When the ore-acid mixture is subsequently baked in the temperature range of 650–700°C, the sulphation reaction continues. The ferric sulphate formed during the pugging stage is unstable at the baking temperature and decomposes according to equation 6. Thus, in both the SO_2 roast and the H_2SO_4 bake processes, the final product is a solid consisting of iron oxide and the sulphates of copper, nickel, cobalt and manganese.

LEACHING PROCESSES

As indicated in Table 9, the various high temperature treatments give rise to a variety of solid products such as alloy phases, metal sulphides, metal chlorides and metal sulphates. The next stage in

the separation process involves the application of hydrometallurgical techniques to these materials in order to dissolve the metal
values and isolate separate metallic products of the individual
valuable metals. The leaching systems that have received the
most attention in the literature on the processing of manganese
nodules are those involving ammonia, chloride and sulphate
solutions. Several of the proposed leaching systems, including
various thermal processing lixiviant combinations, are shown in
Table 10.

Ammoniacal systems

With the exception of the high temperature and the cuprion
processes, all the ammoniacal leaching systems involve the dissolution of metallic phases of copper, nickel and cobalt into their
respective ammonia complexes (called *ammines*) in the presence
of an oxidant, usually air. For a doubly charged (i.e. divalent)
metal ion, the chemical reaction can be described with the following equation where M stands for the divalent metal ion and n for
the oxidation state:

$$M + \tfrac{1}{2}O_2 + nNH_3 + H_2O \longrightarrow M(NH_3)_n^{2+} + 2(OH)^- \qquad (10)$$

Copper, nickel, cobalt, manganese and iron all form divalent
amine complexes, the weakest of which is the ferrous ammine
complex ($n = 2$) followed by the manganese ammine complexes
($1 < n < 4$). In addition, copper forms cuprous ammines ($n = 1$ or 2)
while cobalt forms extremely strong cobaltic ammine complexes
where cobalt is in the $+3$ oxidation state ($n = 3$).[8] It is the relatively high stability of the copper, nickel and cobalt ammine
complexes that permits the selective extraction of these metals
from manganese and iron-containing materials.

During the leaching process, any iron present as a metallic
phase or as ferrous oxide (FeO) dissolves and reprecipitates as
ferric hydroxide. Similarly, any manganous oxide (MnO) produced by the reductive roast operation dissolves and subsequently
precipitates as manganous carbonate ($MnCO_3$). Laboratory experiments[9] as well as practical experience from the laterite processing industry[10] suggests that valuable metals from solution (copper, nickel and cobalt) may be lost during the precipitation of the
ferric hydroxide and manganous carbonate with cobalt showing

Table 10. Pretreatment Leaching Combinations

Lixiviant System	High Temperature Pretreatment	Leaching Method	Development Stage	General Processes
Ammonia	None	Cuprion ammoniacal Leach	Pilot plant	None
		High temperature ammoniacal leach	Laboratory	None
	Reductive roast	Oxidative ammoniacal leach	Pilot plant	Caron process: Cuba, Philippines, Australia, etc.
	Smelt			
	Segregation roast	Oxidative ammoniacal leach	Unknown	Laterites: pilot plant
Chloride	None	Hydrochloric acid leach	Laboratory	Laterites: laboratory
		Low temperature		
		High temperature		
	Chloridizing roast	Water leach	Unknown	Laterites: laboratory
				Manganese ores: laboratory
	Segregation roast	Hydrochloric acid leach	Laboratory	Laterites (laboratory)
	Molten salt chlorination	Water leach	Laboratory	Unknown

Table 10. (continued)

Lixiviant System	High Temperature Pretreatment	Leaching Method	Development Stage	General Processes
Sulphates	None	Low temperature sulphuric acid leach	Laboratory	Oxide copper ores: commercial plant
		High temperature sulphuric acid leach	Pilot plants	Laterites: Moa B
		Sulphur dioxide or sulphuric acid leach	Laboratory	Manganese ores: Laboratory
	Smelting to alloy	Sulphuric acid leach	Laboratory	—
	Smelt to matte	Oxidative sulphuric acid leach	Laboratory	Mattes derived from sulphides or laterites: commercial plants
	Segregation roast	Sulphuric acid		
	Sulphating roast	Water leach		

the greatest degree of vulnerability. In addition, apparently the presence of metallic iron in the roast product can interfere with metal extraction by promoting the formation of air-formed oxide films on the solid surface which cause the passivation of the solid surface.[11]

The cuprion process[12] also permits the selective dissolution of copper, nickel and cobalt leaving iron and manganese behind as insoluble oxides and carbonates. However, unlike the reductive roast-leach process, this novel process developed by Kennecott, specifically for nodule processing, uses an internally generated catalytic reagent (i.e. the cuprous diammine complex) to facilitate the aqueous decomposition of the manganese dioxide phase thereby releasing the contained valuable metals. Equations (11) and (12) below describe the reductive dissolution of MnO_2 that is accompanied by the oxidation of cuprous ammine to cupric ammine and the regeneration of the cuprous ammine respectively:

$$MnO_2 + 2Cu(NH)_2 + 2NH_3 + (NH_4)_2 CO_3 \longrightarrow$$
$$MnCO_3 + 2Cu(NH_3)_4^{2+} + 2(OH)^- \tag{11}$$

$$2Cu(NH_3)_4^{2+} + Co + 2OH^- \longrightarrow$$
$$2Cu(NH_3)_2^+ + 2NH_3 + (NH_4)_2CO_3 \tag{12}$$

The overall reaction is given by equation (13) below.

$$MnO_2 + CO \longrightarrow MnCO_3 \tag{13}$$

This leaching process has been successfully tested by Kennecott on a 350 kg/day ore pilot plant scale.[13]

The high temperature ammoniacal leach is probably based on the ability of ammonia to act as a reductant for tetravalent manganese oxide under the reaction conditions.[14]

$$MnO_2 + 2NH_3 \longrightarrow MnO_3 + 3H_2O + N_2 \tag{14}$$

This process, although also developed by Kennecott, has not been as extensively tested as the cuprion process. Aside from the need for a high temperature pressure leach, a major drawback is the destruction of the ammonia lixiviant; the regeneration of ammonia from nitrogen is not likely to be economic.

Chloride systems

Leaching processes conducted in chloride solutions include those in which nodules are leached directly in hydrochloric acid solutions[15] as well as those in which the leaching operation is preceded by a high temperature chloridization treatment.[16] In the case of the direct leaching methods, both low temperature (100°C) leach and high temperature and pressure leach (greater than 200°C) processes have been investigated. The leaching process chemistry is based on the oxidation of hydrochloric acid by manganese oxide with a corresponding destruction of the oxide matrix and the generation of chlorine:

$$MnO_2 + 4HCl \longrightarrow Mn^{2+} + 2H_2O + 2Cl_2 \tag{15}$$

The available information indicates that the high temperature process results in higher extractions of copper, nickel and cobalt. In both processes, the pregnant solutions contain high concentrations of manganese. However, due to the low solubility of iron at high temperatures, the high temperature leach gives a lower iron dissolution as compared with the low temperature process.

Again, the chloridizing roast yields an intermediate product of copper, nickel, cobalt and manganese chlorides admixed with ferric oxide. A similar product is given by the molten salt chlorination pretreatment. The chlorides are readily soluble in water, although it is found that use of an acidified hydrochloric solution (pH approx. 2.0) is needed to prevent copper hydrolysis. Deepsea Ventures has devoted a considerable amount of research and development effort in the chloridizing roast-leach process including the operation of a pilot plant.[16] The molten salt chlorination process, however, appears to have been limited to bench scale investigations only.

Sulphate systems

A number of direct leach processes involving sulphate media, have been proposed.[17] In the direct leaching systems, manganese dioxide decomposes according to the reaction:

$$MnO_2 + 2H^+ \longrightarrow Mn^{2+} + H_2O + \tfrac{1}{2}O_2 \tag{16}$$

Similarly iron, present in the ore as goethite, dissolves to give ferric sulphate:

$$FeOOH + 3H^+ \longrightarrow Fe^{3+} + H_2O \qquad (17)$$

These reactions release into solution any copper, nickel and cobalt present in the manganese and iron oxide matrix minerals.

The leaching process involving sulphate media is nonselective and extremely slow at low temperatures. On the other hand, the elevated temperature process is relatively fast and gives low iron solubility. The behaviour of iron is due to the fact that ferric ions are unstable at high temperatures and therefore reprecipitate to give hematite (Fe_2O_3).

$$Fe^{3+} + {}^3/_2 H_2O \longrightarrow \tfrac{1}{2} Fe_2O_3 + 3H^+ \qquad (18)$$

The manganese content of the leach liquor can be decreased significantly if an oxygen overpressure is maintained in the leaching reactor. Thus the high temperature, high pressure sulphuric acid leaching method is a selective process for copper, nickel and cobalt extraction.

Another sulphate lixiviant system involves the use of SO_2 as an aqueous phase reductant for converting tetravalent manganese to the divalent state.[18] When this reagent is used, the gas may be bubbled directly into the leaching vessel, or it can be used as sulphurous acid (prepared separately by dissolving SO_2 gas in water). In either case, the reaction with nodules (equation 19) is rapid and high solubilization of copper, nickel, cobalt, manganese and iron is attained.

$$MnO_2 + SO_2 \longrightarrow Mn^{2+} + SO_4^{2-} \qquad (19)$$

According to a recent patent (Agarwal *et al.*), it is possible to arrange process variables (e.g. leaching time, acid/ore ratio) to effect the selective separation of nickel and cobalt from copper.[19]

The use of ferrous ions (Fe^{2+}) as a reductant in sulphate solutions has also been investigated. The chemical reaction responsible for the decomposition of the manganese oxide matrix can probably be described as follows:

$$MnO_2 + 2Fe^{2+} + 4H^+ \longrightarrow Mn^{2+} + 2Fe^{3+} + 2H_2O \qquad (20)$$

SOLUTION TREATMENT

The representative compositions of pregnant leach liquors generated by leaching nodules in either ammoniacal, chloride or sulphate media are presented in Table 11 below.

TABLE 11. Composition (grams/litre) of Representative Nodule Pregnant Leach Solutions

Metal	Ammonia[a]	Ammonia[b]	Chloride[c]	Sulphate[a]	Sulphate[d]	Sulphate[e]
Copper (Cu)	0.2	5.7	7.68	1.2	2.0	3.6
Nickel (Ni)	2.0	6.2	9.54	2.2	3.0	4.3
Cobalt (Co)	1.0	0.2	1.83	0.27	1.0	0.6
Manganese (Mn)	32.0	—	200.0	78.0	73.0	4.5
Iron (Fe)	4.2	—	0.0	4.0	$\leqslant 0.8$	0.3
Zinc (Zn)	—	—	—	—	$\leqslant 0.5$	—
Ammonia (NH_3)	170.0	90.0	—	—	—	—
Carbon dioxide (CO_2)	81.0	55.0	—	—	—	—
Sulphate ions (SO_4^{2-})	—	—	—	66	—	20.0^+
Chloride ions (Cl^-)	—	—	—	—	—	—
(pH)	—	—	2.0	2.0	3.5	—

[a] Brooks, P.T., and Martin, D.A. *Processing Manganiferous Sea Nodules*, U.S. Bureau of Mines, Report of Investigations 7473, 1971.

[b] Agarwal, J.C., *et al.*, A New Fix on Metal Recovery from Sea Nodules, *Engineering and Mining Journal*, December 1976, pp. 74–76.

[c] Cardwell, P.H., Method for separating metal constituents from ocean floor nodules, *U.S. Patent 3,950,486*. 13 April 1976.

[d] Kauczor, H.W., Junghauss, H. and Roever, W., The hydrometallurgy of metalliferous solutions in the processing of nodules, *INTEROCEAN 73*, pp. 469–473.

[e] Neuschutz, D. and U. Scheffer, 1980. Extraction behavior of metal elements fron deep sea manganese nodules in reducing media, *Marine Mining*, **12**, pp. 155–169.

It can be seen from this table that the word *selective* must be taken in a relative sense. Thus for example, even though the sulphuric acid pressure leach process is considered for the recovery of copper, nickel and cobalt, the pregnant liquor nevertheless contains iron and manganese.

The methods that have been proposed for separating the various metals from solution, are based on the nature of the dissolved metals as well as on their absolute and relative concentrations. As shown in Table 12 below, dissolved, copper, nickel, cobalt, manganese and iron exist in ammoniacal solutions as cationic ammine complexes (i.e. positively charged entities in which the metal ions are chemically bonded to ammonia molecules). In contrast, in solutions containing high chloride ion concentrations, with the exception of nickel, all the metals indicated in Table 12 exist as anionic (that is negatively charged) chlorocomplexes. In sulphate solutions, the dissolved metals are present as positively charged ions. Metal separation is much more difficult in sulphate solutions than in either ammoniacal or chloride solutions.

Table 12. Nature of Aqueous Species in Leach Liquors

Pregnant Solutions	Copper	Nickel	Cobalt	Iron	Manganese
Ammoniacal Solutions	$Cu(NH_3)_n^{2+}$ $Cu(NH_3)_n^{+}$	$Ni(NH_3)_n^{2+}$	$CO(NH_3)_n^{3+}$	$Fe(NH_3)_n^{2+}$	$Mn(NH_3)_n^{2+}$
Chloride Solutions	$CuCl_n^{2-n}$	Ni^{2+}	$CoCl_n^{2-n}$	$FeCl_n^{2-n}$	$MnCl_n^{2-n}$
Sulphate Solutions	Cu^{2+}	Ni^{2+}	Co^{2+}	Fe^{3+} Fe^{2+}	Mn^{2+}

A number of flowsheets have been presented in the literature for processing pregnant leach liquors generated by leaching nodules. A few examples are presented in Figs 4 through 11. Figures 4 and 5 show flowsheets for treating ammoniacal liquors; Fig. 6 that for treating chloride liquors and Figs 7 through 11 those for treating sulphate liquors.

Ammoniacal Solutions

Figures 4 and 5 show flowsheets for treating ammoniacal liquors.

Figure 4 proposed by Brooks and Martin is based on a series of selective precipitation steps.[20] Firstly, manganese is removed as a carbonate by taking advantage of the relatively low thermal stability of manganese amine complexes. After removing copper and iron as sulphides and hydroxide precipitates respectively, the resulting leach liquor is boiled to give a mixed nickel – cobalt carbonate product while ammonia and carbon dioxide are driven into the vapour phase and recycled to the leaching state. The mixed carbonate can be processed by conventional technology.

An alternative process for treating ammoniacal carbonate pregnant liquors is presented in Fig. 5. This process, a solvent

Fig. 4. Treatment of Ammoniacal Carbonate Leach Liquors by Precipitation
Techniques (After Brooks and Martin)

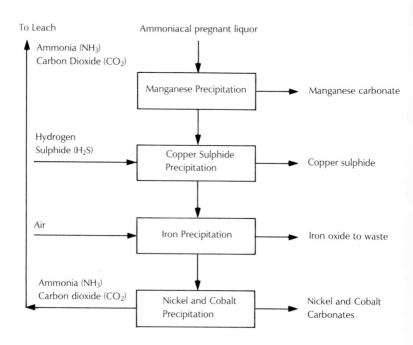

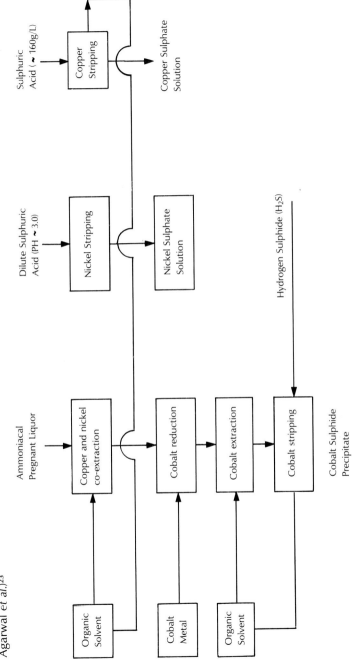

Fig. 5. Treatment of ammoniacal leach liquors by solvent extraction (After Agarwal *et al.*)[23]

extraction based technology, was developed by Kennecott.[21] By maintaining cobalt in the oxidation state of plus three, copper and nickel are selectively co-extracted into an organic phase containing a selective chelating reagent (LIX64N). Selective stripping of the loaded organic phase with sulphuric acid generates separate aqueous solutions of nickel and copper sulphate from which the metals can be recovered through electrowinning.[22] The aqueous solution, now depleted of copper and nickel ions is subsequently treated with metallic cobalt to reduce cobalt from the plus three (Co^{3+}) to the plus two (Co^{2+}) oxidation state whereupon it becomes extractable by the solvent extraction reagent (LIX64N). Cobalt is stripped from the organic phase as a cobalt sulphide with the aid of hydrogen or ammonium sulphide. The cobalt sulphide can then be processed by conventional technology.

Chloride Solutions

Most of the work that has been conducted on chloride systems has been done by the Deepsea Ventures consortium. Figure 6 shows a simplified flowsheet of one of their solution treatment procedures.[23] Copper is first removed at a low pH (2) by a selective chelating extractant (e.g. KELEX 100).[24] Using the same extractant, nickel and cobalt are co-extracted at a higher pH (4–5) leaving behind a manganese chloride solution. Selective stripping of the loaded organic solution gives separate nickel and cobalt chloride solutions.[25]

Sulphate Solutions

Figures 7 through 11 present five different flowsheets that have been presented in the literature for treating sulphate liquors generated by nodule leaching. Figure 7, based on the work of Kauczor *et al.*, uses a combination of ion exchange and solvent extraction.[26] Copper, nickel, cobalt, iron and zinc are selectively removed from manganese by adsorbing them on an ion exchange resin (Lewatit TP 207). Copper is the metal that is most strongly bound to the resin and, therefore, nickel, cobalt and zinc are selectively eluted with dilute hydrochloric acid. The copper-loaded resin is then eluted with sulphuric acid to give a copper sulphate solution. After increasing the chloride content of the first eluate with NaCl, zinc is then selectively extracted, followed by cobalt with an amine extractant.

Fig. 6. Treatment of chloride leach liquors by solvent extraction (After Cardwell, 1976)[27]

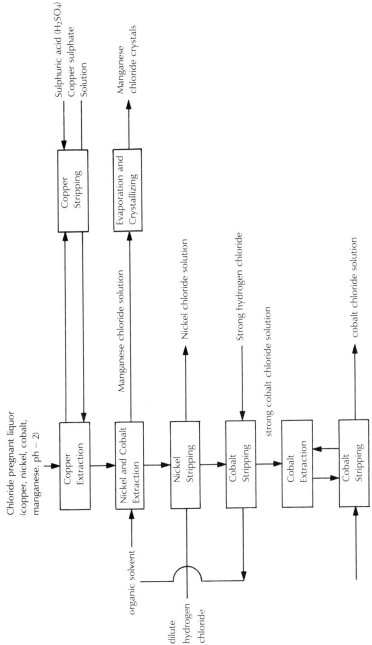

Brooks and Martin have proposed the flowsheet shown in Fig. 8 for treating solutions generated by the sulphating roast-leach process. An impure metallic copper product is produced by cementing out dissolved copper with metallic iron.[27] After removing nickel and cobalt as a mixed sulphide, the remaining manganese sulphate solution is subjected to a high temperature (200°C) treatment to yield manganese sulphate crystals.

A solvent extraction procedure that is suitable for treating pregnant solutions generated by the sulphuric acid pressure leach process is presented in Fig. 9. According to this procedure that was developed by Agarwal *et al.,* copper is first selectively extracted at a pH of about 2.[28] After neutralizing the solution with ammonia, nickel and cobalt are then co-extracted and selectively stripped to give their respective sulphate solutions. Two alternative procedures and their corresponding flowsheets (Figs 10 and 11) have been proposed by Neuschutz and Scheffler for sulphuric acid pressure leach liquors.[29] In one of these procedures (Fig. 10), copper is first removed by solvent extraction in the form of a concentrated pure cupric sulphate solution. Nickel and cobalt are then co-extracted from the copper barren rafinate by ion exchange. The remaining sulphate solution, containing manganese and iron, is subjected to oxyhydrolysis to remove these metals as a hydroxide precipitate. After eluting the loaded resin with a dilute solution of hydrochloric acid, cobalt and nickel are selectively precipitated as cobalt and nickel hydroxides.

In the second procedure, the ion exchange – precipitation steps for nickel and cobalt are replaced by sulphide precipitation.

INTERMEDIATE PRODUCTS

Table 13 presents a summary of the various metal-containing products that are referred to in the flowsheets shown in Figs 4 to 11. In the case of copper, the preferred separation method among the eight shown is *solvent extraction* which is incorporated in five of the flowsheets (Figs 5, 6, 9, 10 and 11). In respect of intermediate copper-containing products, *cupric sulphate solution* is produced in six out of the eight flowsheets (Figs 5, 6, 7, 9, 10 and 11).

In the case of *nickel,* the use of *solvent extraction techniques* is indicated in three of the flowsheets (Figs 5, 6 and 9) with a balance between ion exchange and precipitation for the remainder. In

Fig. 7. Combined Ion Exchange – Solvent Extraction Process for Treating Sulphate Leach Liquors (After Kauczor *et al.* 1973)[28]

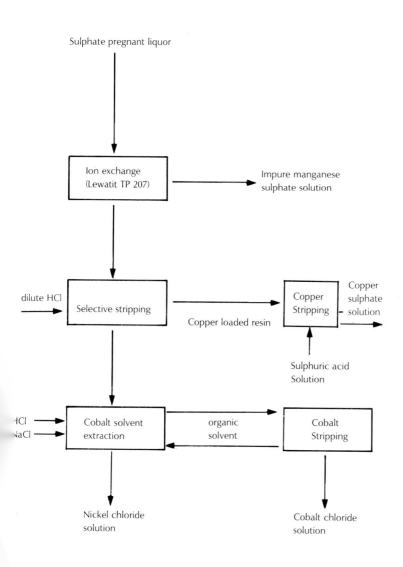

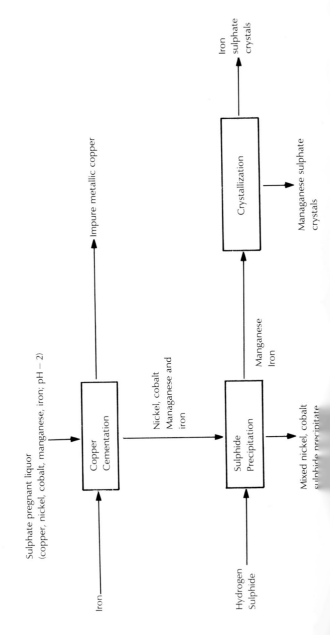

Fig. 8. Precipitation-crystallization process for treating sulphur dioxide leach or sulphating-roast-leach liquors (After Brooks and Martin 1971)[29]

Fig. 9. Solvent extraction process for treating sulphuric acid pressure leach liquors (After Agarwal)[30]

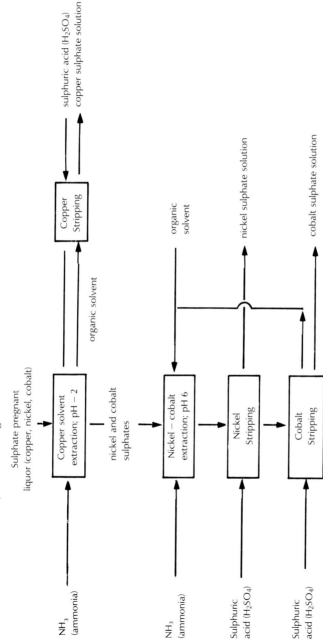

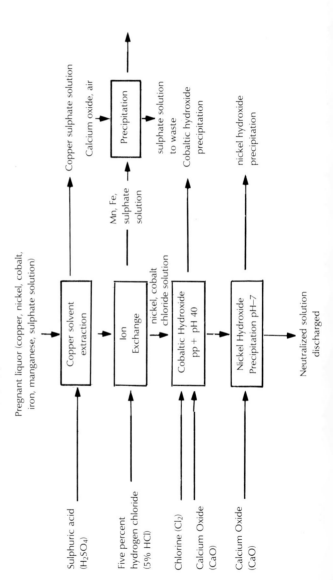

Fig. 10. Combined solvent extraction – ion exchange – selective precipitation method for treating sulphuric acid pressure leach pregnant liquors (Neuschutz and Scheffler)[31]

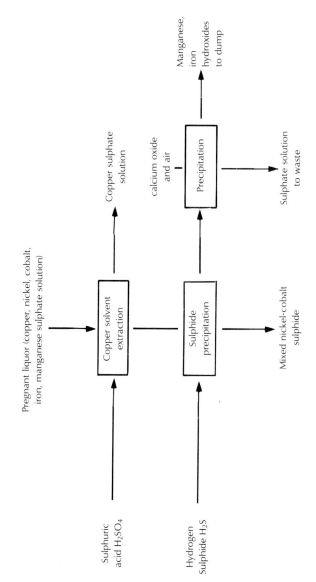

Fig. 11. Combined solvent extraction – selective precipitation method for treating sulphuric acid pressure leach pregnant liquors (After Neuschutz and Scheffler)[31]

respect of intermediate nickel-containing products, two of the process flowsheets generate a *nickel sulphate solution* (Figs 5 and 9), two generate *a mixed nickel-cobalt sulphide precipitate* (Figs 8 and 11), another two produce *a nickel chloride solution* (Figs 6 and 7) while the remaining two flowsheets yield *a mixed nickel-cobalt carbonate precipitate* (Fig. 4) and *a nickel hydroxide precipitate* (Fig. 10) respectively.

For *cobalt,* solvent extraction techniques by themselves account for three out of the eight flowsheets (Figs 5, 6 and 9). In respect of intermediate cobalt-containing products, two of the process schemes generate *a mixed nickel-cobalt sulphide precipitate* (Figs 8 and 11), two generate *a cobalt chloride solution* (Figs 6 and 7) while the remaining four flowsheets produce *a mixed nickel-cobalt carbonate precipitate* (Fig. 4), a *cobalt sulphide precipitate* (Fig. 5), a *cobalt sulphate solution* (Fig. 9) and a *cobaltic hydroxide precipitate* (Fig. 10) respectively.

A major advantage of solvent extraction is that it can be selective and can give rise to a purified and concentrated solution of a given valuable metal that can be used in direct metal production (by electrowinning for example). Precipitation methods lead to impure products that have to be further refined through such methods as redissolution and solvent extraction. However, since the cost of a solvent extraction operation (both capital and operating costs) may be considered proportional to the volume of aqueous solution that is processed, in some cases, the uses of solvent extraction following the redissolution of a small amount of precipitate is to be preferred to the case where the original voluminous pregnant solution is subjected to solvent extraction.

METAL AND REAGENT RECOVERY

Once the one-metal aqueous solutions or two-metal precipitates have been derived from the separation methods discussed above, the next step is to produce a marketable product. It is conceivable that where a solid intermediate product such as a mixed nickel-cobalt sulphide is formed, a commercial nodule processing plant may decide to sell this material rather than refine it in-house. However, even in this case, the company that receives such an intermediate product will still have to subject it to a refining operation.

Table 13. Metal Separation Methods

Nature of Solution	Copper	Nickel	Cobalt	Manganese
1. Ammonia (Fig. 4; After Brooks and Martin)	A. Precipitation B. H_2S C. Copper sulphide ppt	A. Precipitation B. Heat (100°C) Mixed Ni-Co C. Carbonate ppt	A. Precipitation B. Heat (100°C) Mixed Ni-Co C. Carbonate ppt	A. Precipitation B. Heat (65°C) Manganese C. Carbonate ppt
2. Ammonia (Fig. 5; After Agarwal et al.)	A. Solvent extraction B. LIX64N, H_2SO_4 C. $CuSO_4$ solution	A. Solvent extraction B. LIX64N, H_2SO_4 C. $NiSO_4$ solution	A. Solvent extraction B. LIX64N, H_2S C. CoS precipitate	A. — B. — C. —
3. Chloride (Fig 6;After Kane and Cardwell)	A. Solvent extraction B. Kelex 100, H_2SO_4 C. $CuSO_4$ solution	A. Solvent extraction B. Kelex 100, HCl C. $NiCl_2$ solution	A. Solvent extraction B. Kelex 100, HCl C. $CoCl_2$ solution	A. Crystallization B. Heat (200°C) C. $MnCl_2$ crystallization
4. Sulphate (Fig 7: After Kauczor)	A. Ion exchange/solvent extraction B. Lewatit TP207, H_2SO_4 C. $CuSO_4$ solution	A. Ion exchange/solvent extraction B. Lewatit TP207, HCl C. $NiCl_2$ solution	A. Ion exchange/solvent extraction B. Lewatit TP207, HCl C. $CoCl_2$ solution	A. Raffinate from Co-Ni-Cu and ion exchange B. — C. $MnSO_4$ solution
5. Sulphate (Fig 8: After Brooks and Martin)	A. Cementation B. Iron, H_2S C. Impure metallic copper	A. Precipitation B. H_2S Mixed nickel C. Cobalt sulphide ppt	A. Precipitation B. H_2S Mixed nickel C. Cobalt sulphide ppt	A. Raffinate from Cu-Ni-Co ion exchange (crystallization) B. — C. $MnSO_4$ crystallization
6. Sulphate (Fig. 9: After Agarwal)	A. Solvent extraction B. LIX64N, H_2SO_4 C. $CuSO_4$ solution	A. Solvent extraction B. LIX64N, NH_3, H_2SO_4 C. $NiSO_4$ solution	A. Solvent extraction B. LIX64N, NH_3, H_2SO_4 C. $CoSO_4$ solution	A. Crystallization B. Heat (200°C) C. $MnSO_4$ crystallization

Table 13. (continued)

Nature of Solution	Copper	Nickel	Cobalt	Manganese
7. Sulphate (Fig. 10: After Neuschutz and Scheffler)	A. Solvent extraction B. LIX64N, H_2SO_4 C. $CuSO_4$ solution	A. Ion exchange/ precipitation B. Chelating resin/CaO C. Ni hydroxide ppt	A. Ion exchange/ precipitation B. Chelating resin/Cl_2, CaO C. Co hydroxide ppt	A. Oxyhydrolysis precipitation B. — C. Mn hydroxide ppt
8. Sulphate (Fig. 11: After Neuschutz and Scheffler)	A. Solvent extraction B. LIX64N, H_2SO_4 C. $CuSO_4$ solution	A. Precipitation B. LIX64N, H_2S C. Mixed Ni-Co sulphide precipitate	A. Precipitation B. LIX64N, H_2S C. Mixed Ni-Co sulphide precipitate	A. Precipitation B. CaO, Air C. Mixed Fe-Mn hydroxide

A. — Method;   B. — Reagent;   C. — Product.

The further the processing operation has been carried beyond raw unprocessed nodules, the lesser the need for developing entirely new separation processes. Thus, once a nodule processing scheme generates an intermediate product such as a cupric sulphate solution or a mixed nickel-cobalt sulphide, conventional technology is available for the preparation of a saleable metallic product. Table 14 presents a summary of current commercial methods for treating the solutions and precipitates generated by the separation operations discussed above. For producing a metallic copper product from purified cupric sulphate solutions, the preferred method, as indicated in the table, is electrowinning. On the other hand, in the case of nickel or cobalt sulphate solutions, both electrowinning and hydrogen reduction methods are utilized to produce the metallic product.

Table 14. Metal Production Methods

Feed Material	Method	Development Status
1. $CuSO_4$ solution	Electrowinning	Conventional technology[a]
2. $NiSO_4$ solution	(a) Electrowinning	(a) Conventional technology[b]
		(i) Outokumpo, Finland
	(b) Hydrogen reduction	(b) Conventional technology[c]
		(i) Sherritt Gordon, Fort Saskatchewan, Canada
		(ii) Amax, Port Nickel, USA
3. $CoSO_4$ solution	(a) Electrowinning	(a) Conventional technology[d]
		(i) Gecamines, Zaire (Shituru)
		(ii) Gecamines, Zaire (Luilu)
		(iii) ZCCM, Zambia (Chambisi)
		(iv) ZCCM, Zambia (Rhokana)
		(v) Nippon Mines, Hitachi, Japan
	(b) Hydrogen reduction	(b) Conventional technology[e]
		(i) Sherritt Gordon, Canada
		(ii) Outokumpo, Finland (Kokkola Plant)
		(iii) Amax, Port Nickel, USA
4. $NiCl_2$ solution	(a) Electrowinning	(a) Relatively new technology[f]
		(i) Falconbridge (mixed chloride-sulphate electrolyte)
		(ii) Societé le Nickel, Sandonville plant, France[g]
		(iii) Sumitomo, Niihama refinery, Japan
	(b) Pyrohydrolysis	(b) Relatively new technology[h]
		(i) Falconbridge
		(ii) Amax (pilot plant)

Table 14. (continued)

Feed Material	Method	Development Status
5. CoCl₂ solution	Electrowinning	Relatively new technology[j] (i) Falconbridge, Kristiansand, Norway (ii) Sumitomo, Niihama refinery, Japan (iii) U.S.S.R.
6. Mixed Ni-Co sulphide	(a) H₂SO₄ pressure leach and H₂ reduction in ammoniacal solution	(a) Conventional technology (i) Sherritt Gordon (ii) Amax, Port Nickel, USA (iii) Outokumpo, Kokkola plant, Finland
	(b) H₂SO₄ pressure leach, solution extraction and crystallization	(b) Matthey-Rusenburs[j]
	(c) H₂SO₄ pressure leach, solvent extraction and electrowinning	(c) Relatively new technology[k] (i) Nippon mining, Hitachi, Japan (ii) Sumitomo, Niihama
	(d) Chloride leach and solvent extraction	(d) Societé le Nickel, Sandonville plant, France
	(e) Chloride leach, solvent extraction and electrowinning	(e) Falconbridge, Kristiansand, Norway
7. Mixed Ni-Co Carbonate	(a) Calcination and gaseous reduction of basic nickel carbonate	(a) Conventional technology[l] (i) Nicaro (ii) Yabulu refinery
	(b) (NH₄)₂SO₄ leach and hydrogen reduction	(b) Conventional technology (i) Marindugue, Surigao refinery
	(c) Calcination	(c) Conventional technology (i) Inco, Copper Cliff: Canada

Table 14. (continued)

Feed Material	Method	Development Status
8. Cobaltic hydroxide	(a) H_2SO_4 reductive leach and hydrogen reduction	Amax, Port Nickel[m]
	(b) H_2SO_4 reductive leach and solvent extraction	Matthey-Rostenberg[n]
	(c) H_2SO_4 reductive leach and electrorefining	Inco, Port Colborne[o]

[a] Biswas, A.K. and W.G. Davenport, *Extractive Metallurgy of Copper*, Second Edition, Pergamon, Oxford.

[b] Queneau, P. ed. 1961. *Extractive Metallurgy of Copper, Nickel and Cobalt*, Interscience, New York.

[c] Boldt, J.R., Jnr. and P.Queneau, 1967. *The Winning of Nickel*, Van Nostrand, Princeton, New Jersey; Lenoir, P., A. Van Peteghem and C. Feneau, Extractive Metallurgy of Cobalt, in *International Conference on Cobalt: Metallurgy and uses*, Vol. 1, Benelux Metallurgie, Brussels, Belgium, pp. 51–62.

[d] Lenoir, P., A. Van Peteghem and C. Feneau, *op. cit.*

[e] Boldt, J.R. Jnr., and P. Queneau *op. cit.*; Lenoir, P., A. Van Peteghem and C. Feneau, *op. cit.*

[f] Lenoir, P., A. Van Peteghem and C. Feneau, *ibid*; Mackinnon, D.J., 1982. The electrowinning of metals from aqueous chloride in K. Osseo-Asare and J.D. Miller eds., *Hydrometallurgy-Research Development and Plant Practice*, pp. 659–677, TMS-AIME, Warrendale, PA.

[g] Gilpin, W.C. and N. Heasman, 1977. Recovery of Magnesium compounds from sea water, *Chemical Industry*, 16 July, pp. 567–572.

[h] Jha, M.C. and G.A. Meyer, AMAX chloride refining process for the recovery of nickel and cobalt from mixed sulphide precipitates, in K. Osseo-Asare and J.D. Miller, eds., *op. cit.*, pp. 903–924, *op. cit.*

[i] Mackinnon, D.J., *op. cit.*

[j] Lenoir, P., *et al.*, *op. cit.*

[k] Boldt, J.R. Jnr. *et al.*, *op. cit.* Crynojevich, R., D.H. Wilkinson and J.L. Blanco, Contributions to the hydrometallurgy of nickel, copper and cobalt in the Amax Inc., Port Nickel Plant in Osseo-Asare and J.D. Miller, eds., pp. 947–953, *op. cit.*

[l] *Ibid.*

[m] Lenoir, P. *et al.*, *op cit.*; Crynojevich, R., D.H. Wilkinson and J.L. Blanco, *op. cit.*

[n] Lenoir, P., *et al.*, *op. cit.*

[o] Renzoni, L.S. and W.V. Barker, Production of electrolytic cobalt at INCO's Port Colbourne Nickel Refinery, in P. Queneau, ed., pp. 535–545, *op. cit.*

NOTES

1. Osseo-Asare, K., 1983. Processing of low-grade oxide ores: Physical and Chemical Constraints, paper presented at the *3rd Symposium on Separation Science and Technology*. Gatlinburg, Tenn., USA, 28 June-1 July.
2. (a) Redman, M.J., 1973. Extraction of metal values from complex ores, *US Patent 3,734,715, 22 May.*
   (b) Wilder, T.C. and J.J. Andreola, 1973. Recovery of copper, nickel, cobalt and molybdenum from complex ores, *US Patent 3, 753, 686,* 21 Aug.
   (c) Brooks, P.T. and D.A. Martin, 1971. *Processing manganiferous sea nodules,* US Bureau of Mines Report of Investigations 7473.
   (d) Han, K.N., M. Hoover and D.W. Fuerstenau, 1974. Ammonia – ammonium leaching of deep-sea manganese nodules, *International Journal of Mineral Processing,* 1, pp. 215–230.
3. (a) Vasil'chikov, N.V., G.B. Shiner, Y.A. Matsepon, I.F. Krasnykh and E.A. Grishankova, 1968. Iron-manganese nodules from the ocean floor raw materials for the production of cobalt, nickel, manganese and copper. *Tsvet. Metal,* 41(1), pp. 40–2.
   (b) Beck, R.R. and M.E. Messner, 1970. Copper, nickel, cobalt and molybdenum recovery from deep sea nodules, in R.P. Ehrlich, ed., *Copper Metallurgy,* AIME, New York.
   (c) Sridhar, R., 1974. Thermal upgrading of sea nodules, *Journal of Metals,* December 1974, pp. 18–22.
   (d) Sridhar, R., W.E. Jones and J.S. Warner, 1976. Extraction of copper, nickel and cobalt from sea nodules, *Journal of Metals,* April 1976, pp. 32–37.
4. Sridhar, R., W.E. Jones and J.S. Warner, Extraction of copper, nickel, and cobalt from sea nodules, *op. cit.*
5. (a) Lenoir, P., A. Van Peteghem and C. Feneau, 1981. Extractive Metallurgy of Cobalt, in *International Conference on Cobalt: Metallurgy and Uses,* Vol. 1, Benelux Metallurgie, Brussels, Belgium, pp. 51–62.
   (b) Queneau, P., ed., 1961. *Extractive metallurgy of copper, nickel and cobalt.* Interscience, New York.
6. (a) Kane, W.S. and P.H. Cardwell, 1973. Recovery of metal values from ocean floor nodules, *US Patent 3,752,745,* 14 August.
   (b) Cardwell, P.H., 1976. Method for separating metal constituents from ocean floor nodules, *US Patent 3,950,486,* 13 April.
   (c) Hoover, M., K.N. Han and D.W. Fuerstenau, 1975. Segregation roasting of nickel, copper and cobalt from deep-sea manganese nodules, *International Journal of Mineral Processing,* 2, pp. 173–185.
   (d) Ngaya, Y., K. Abe and T. Masaru, 1976. *Japan Patent 76,96,724.*
7. (a) Iwasaki, I., A.S. Malicsi and N.C. Jagolino, Segregation process for copper and nickel ores, *Progress in Extractive Metallurgy,* Vol. 1.
   (b) Meyer-Galow, E., K.H. Schwarz and U. Boin, Metal extraction from manganese nodules by sulphating treatment, in *INTEROCEAN '73,* Vol. 1, pp. 458–468.
   (c) Van Hecke, M.C. and R.W. Bartlett, 1973. Kinetics of sulphation of Atlantic ocean manganese nodules, *Metall. Trans.,* 4, pp. 941–947.
   (d) Kane, W.S. and P.H. Cardwell, 1975. Reduction method for separating metal values from ocean floor nodule ore, *US Patent 3,869,360,* 4 March.
   (e) Baba, R. and T. Tamagawa, 1979. *Nippon Kogyo Kaishi,* 95, (1096), 353.

8. (a) Lee, J.J., J.W. Gilje, H. Zeitlin and Q. Fernando, 1980. *Separation Science Technology*, **15**.

   (b) Lewis, W.E., L.F. Heising, J.W. Pennington and C. Prasky, 1958. *Investigation of cuyuna iron-range manganese deposits.* US Bureau of Mines, Department of Interior Report of Investigations, 5400.

   (c) Joyce, F.E., Jnr., 1965. *Sulphatization of nickeliferous laterites*, US Bureau of Mines, Rep. of Investigations 6644.

   (d) Zubryckyj, N., D.J. Evans and V.N. Mackiw. Preferential sulphation of nickel and cobalt in lateritic ores, *Journal of Metals*, May 1965, pp. 478–486.

9. (a) Osseo-Asare, K. and D.W. Fuerstenau, 1978. Application of activity-activity diagrams to ammonic hydrometallurgy: The systems Cu– NH_3–H_2O, Ni – NH_3 – H_2O and Co – NH_3 – H_2O at 25°C, in T.W. Chapman, *et al.*, eds, *Fundamental Aspects of Hydrometallurgical Processes*, AIChE Symposium series, Vol. 74 (173), pp. 1-13.

   (b) Osseo-Asare, K., 1981. Application of activity-activity diagrams to ammonia hydrometallurgy III. The Mn – NH_3 – H_2O, Mn – NH_3 – H_2O – CO_3 and Mn – NH_3 – H_2O – SO_4 systems at 25°C, *Transactions of the Institute of Mining and Metallurgy*, 90C, pp. 152–158.

   (c) Osseo-Asare, K. and S.W. Asihene, 1979. Hetrogeneous equilibra in ammonia/laterite leaching systems in D.J.I. Evans *et al.*, eds, *International Laterite Symposium*, AIME, New York, pp. 585–609.

   (d) Osseo-Asare, K., 1981. Applicaton of activity-activity diagrams to ammonia hydrometallurgy IV. The Fe – NH_3 – H_2O, Fe – NH_3 – H_2O – CO_3 and Fe – NH_3 – H_2O – SO_4 systems at 25°C, *Transactions of the Institute of Mining and Metallurgy*, 90C, pp. 159–163.

10. (a) Osseo-Asare, K. and D.W. Fuerstenau, 1978. Adsorption losses in ammonia leaching of copper, nickel and cobalt from deep sea manganese nodules, in M.J. Jones, ed., *Complex Metallurgy, 1978*, The Institute of Mining and Metallurgy, London, pp. 43–48.

    (b) Han, K.N., E. Narita and F. Lawson, 1982. The coprecipitation behaviour of Co(II) and Ni(II), with Fe(II), Cr(III) and Al(III) from aqueous ammoniacal solutions, *Hydrometallurgy*, **8**, pp. 365–377.

11. Weir, D.R. and V.B. Sefton, Development of Sherritt's Commercial nickel refining process for low and high iron laterites, in D.J.I. Evans *et al.*, eds, *op cit.*

12. (a) Reid, J.G., 1982. Some observations of roasting, leaching and washing characteristics of Greenvale lateritic ore, in *Hydrometallurgy-Research Development and Plant Practice*, pp. 109–120.

    (b) Goeller, L.A., C.D. Low and D.C. Gale, The recovery of cobalt from laterite ore, *International Conference on Cobalt.* pp. 85–105.

13. Goeller, L.A., C.D. Low and D.C. Gale, *op. cit.*

14. Osseo-Asare, K., J.W. Lee, H.S. Kim and H.W. Pickering, 1983. Cobalt extraction in ammoniacal solution: Electrochemical effects of metallic iron, *Meta. Trans. Bull.*, **14**, pp. 571–576.

15. (a) Gong, Q., 1980. *Non-ferrous Metals Quarterly (China)*, **32**, (2).

    (b) Queneau, P., ed., *Extractive Metallurgy of Copper, Nickel and Cobalt, op cit.*

16. Lenoir, P., A. Van Peteghem and C. Feneau, 1981. Extractive Metallurgy c

Cobalt in *International Conference on Cobalt: Metallurgy and Uses,* Vol. 1. Benelux Metallurgic, Brussels, Belgium.

(a) Cardwell, P.H., Extractive Metallurgy of Manganese Nodules, *Mining Congress Journal*, November 1973, pp. 38–43.

(b) Lenoir, P. *et al.*, Extractive metallurgy of Cobalt, *op. cit.*

(c) Queneau, P. *Extractive Metallurgy of Copper, Nickel and Cobalt, op. cit.*

17. (a) Szabo, L.J., 1976. Recovery of metal values from manganese deep sea nodules using ammoniacal cuprous leach solutions, *US Patent 3,983,017*, 28 September.

(b) Agarwal., J.C., H.E. Barner, N. Beechner, D.S. Davies and R.N. Kust, Kennecott process for recovery of copper, nickel, cobalt and molybdenum from ocean nodules, *Mining Engineering*, December 1979, pp. 1704–1707.

(c) Skarbo, R.R., 1973. Extraction of metal values from manganese deep sea nodules, *US Patent 3,728,105*, 17 April.

(d) Ulrich, K.H., U. Scheffler and M.J. Meixner, 1973. The processing of nodules by acid leaching, *INTEROCEAN '73,* 1.

18. (a) Neuschutz, D. and U. Scheffler, 1976. Processing of manganese nodules by sulphuric acid pressure leaching, *INTEROCEAN '76*, pp. 102–114.

(b) Han, K.N. and D.W. Fuerstenau, 1980. Extraction behaviour of metal elements from deep-sea manganese nodules in reducing media, *Marine Mining*, **2**, pp. 155–169.

(c) Pahlman, J.E. and S.E. Khalafalla, 1979. Selective recovery of nickel, cobalt and managanese from sea nodules with sulphurous acid, *US Patent* 4,138,465, 6 Feb.

(d) Kane, W.S. and P.H. Cardwell, 1975. Winning of metal values from ore utilizing recycled acid leaching agent, *US Patent* 3,923,615, 2 Dec.

(e) Agarwal, J.C. *et al.*, A new fix on metal recovery from sea nodules, *Engineering and Mining Journal*, December 1976. pp. 74–76.

19. Agarwal, J.C. *et al.*, *op. cit.*

20. Brooks, P.T. and D.A. Martin, Processing manganiferous sea nodules, *op. cit.*

21. (a) Agarwal, J.C. *et al.*, A new fix on metal recovery from sea nodules, *Engineering and Mining Journal*, December 1976, pp. 74–76.

(b) Hubred, G.L., R.N. Kust, D.L. Natwig and J.P. Pemsler, 1978. Cobalt stripping process, *US Patent 4,083,915*. 11 April.

22. Electrowinning is the recovery of metals from solution by electrolysis.

23. Mackinnon, D.J., 1982. The electrowinning of metals from aqueous chloride in K. Osseo-Asare and J.D. Miller, eds, *Hydrometallurgy-Research, Development and plant practice*, TMS-AIME, Warrendale, Pa, pp. 659–677.

24. pH — This represents the negative logarithm of the effective hydrogen-ion concentration or hydrogen-ion activity in grain equivalents per liter used in expressing both acidity and alkalinity on a scale whose values run from 0 to 14. 7 represents neutrality, numbers less than 7 increasing acidity and numbers greater than 7 increasing in alkalinity.

25. Cardwell, P.H., 1976. Method for separating metal constituents from ocean floor nodules, *US Patent 3,950,486*, 13 April.

26. Kauczor, H.W., H. Junghanss and W. Roever, 1973. The hydrometallurgy of metalliferous solutions in the processing of manganese nodules, *INTER-OCEAN '73*, pp. 469–473.

27. Wilder, T.C. and J.J. Andreola, 1973. Recovery of copper, nickel, cobalt and molybdenum from complex ores, *US Patent 3,753,686*, 21 August.

28. Agarwal, J.C., N. Beecher, D.S. Davis, G.L. Hubred, V.K. Kakaria and R.N. Kust, Processing of ocean nodules: A technical and economic review, *Journal of Metals*, April 1976, pp. 24–31.
29. Neuschutz, D. and U. Scheffler, 1976. Processing of manganese nodules by sulphuric acid pressure leaching, *INTEROCEAN '76*, pp. 102–114.

Chapter 6

Metallurgical Work Undertaken by the Various Consortia[1]

Of the variety of processing schemes for recovering the valuable metals in nodules, significant research has been undertaken in the development of four processes. These processes are:

(i)   the Kennecott cuprion process;
(ii) the Inco process (a combination of pyrometallurgical and hydrometallurgical methods);
(iii) the Deepsea Ventures (Ocean Mining Associates) processes, recovering manganese as well as nickel, copper and cobalt and,
(iv) the high temperature sulphuric acid leach.

THE KENNECOTT CUPRION PROCESS

Kennecott Corporation's cuprion process, like the Caron nickel process being operated in Nicaro, Cuba, is based on ammonium carbonate leaching of nickel. In the caron process, laterite ore is dried and pre-reduced at high temperatures with gases. The cuprion process was developed to eliminate this energy-intensive stage and instead to carry out the reduction during leaching by means of cuprous ions.

In the process, wet ore is ground and then slurried in a mixture of sea water and recycled process liquor which contains dissolved copper and ammoniacal ammonium carbonate. The slurry passes through a series of reaction vessels into which carbon monoxide is introduced. Cuprous ions are produced which subsequently catalyze the reduction of the manganese iron oxide matrix.[2] Valued metals dissolve and are separated from the reduced residues by countercurrent washing. Ammonia and carbon dioxide are recovered and recycled by steam stripping residues. Electro-won nickel and copper are extracted from the leach liquor using a mixture of LIX64N in Kerosine.[3] By precipitating the raffinate with hydrogen sulphide (H_2S), cobalt is recovered in the form of an impure sulphide which forms part of the feed to a separate cobalt extraction circuit. This precipitate can be treated in differ-ent ways to recover the cobalt. Though the processes may vary in detail, typically they would include a re-leaching of the sulphide, solvent extraction to remove copper and nickel, as well as zinc and molybdenum if present, and electrowinning of the cobalt from the raffinate.

Kennecott has not indicated any intention of recovering manga-nese from leach tailings. However, flotation testing by Kennecott reportedly produced manganese concentrates of commercial grade and quality from leach tailings. If commercially applicable, such a flotation step would enable the sale of manganese concen-trates or ferromanganese smelting on site.

The Kennecott cuprion process has a number of factors to recommend it.[4] First, all the steps in the process take place at ambient temperature and pressure. Second, energy consumption is relatively low. Third, most of the reagents used in the process are relatively inexpensive or recyclable and fourth, there is only limited use of corrosive and highly toxic reagents.

THE INCO PROCESS

Inco's process is a combination of pyrometallurgical and hydro-metallurgical methods that have been used in the processing of land-based ores. In the process, nodules, containing about 30 percent water, are dried and then selectively reduced at a tempera-ture of 1400°C. A manganese-rich slag and an iron-nickel-cobalt-copper alloy are produced. The next step in the process is the oxidization of the alloy to remove most of the iron and manganese and conversion of the alloy to a sulphide matte by the addition of pyrite, gypsum and coke. The matte is then reoxidized (blown) to remove residual iron. The slag is returned to the reduction step. The remaining matte is ground and pressure leached with sulphuric acid. Metals are recovered by solvent extraction, elec-trowinning and precipitation. Ferromanganese with acceptable manganese to phosphorous ratios is produced from the manganese slag by reduction smelting with lime at a temperature of 1600°C.

Factors that contribute to recommending this process include, first, nearly all the manganese and much of the iron contained in nodules are removed in a molten slag which can be used to produce ferromanganese; second, other valuable commodities are concentrated into an alloy phase which weighs less than 10 percent of the original feed; third, the treatment of this smaller quantity of material is relatively cheap compared with other processes; fourth, most of the commercial technology already exists.[5] The principal disadvantage to this process is the high energy requirements for drying and smelting.

THE DEEPSEA VENTURES (OCEAN MINING ASSOCIATES) PROCESS

This process is based on the reaction of the manganese–iron hydroxide matrix in nodules with hydrochloric acid (HCl). The reaction produces soluble iron and manganese chlorides thereby releasing the contained metal values. Ferric chloride is removed by solvent extraction and oxidized to produce recyclable hydrochloric acid (HCl) and iron oxide. Copper, and then nickel, cobalt and molybdenum are separated by liquid ion exchange (LIX) from the chloride solution and electrowon in separate chloride circuits. Manganese metal is recovered by either fused salt electrolysis or reduction with aluminium metal. Alternately, manganese oxide can be recovered by high-temperature hydrolysis of manganese chloride ($MnCl_2$).

The number and diversity of patents issued on the process appear to suggest that problems have been experienced with it. A major problem is the large consumption of hydrochloric acid during the initial reduction and the accompanying production of chlorine which is not utilized in the remainder of the process.[6] Another problem, although minor, is the development of a satisfactory method to convert the manganese oxide intermediate product to a usable end product such as ferromanganese. A recent publication suggests that with the addition of Metallurgie Hoboken-Overpelt to the consortium through its corporate ties with Union Miniere of Belgium, some of these problems may be resolved.[7]

THE HIGH TEMPERATURE SULPHURIC ACID LEACH PROCESS

Of the various processes described in the literature, the high temperature sulphuric acid leach process is the one on which the least amount of information is available. This process has, however, been investigated in European laboratories and is an adaptation of the technique used to recover nickel from laterite ores at Moa Bay, Cuba.[8]

Raw, wet nodules are ground, mixed with sulphuric acid and heated to about 250°C. Most of copper, nickel and cobalt are dissolved at this temperature, while little iron or manganese enter into solution. As a result, subsequent purification steps are simplified and the consumption of acid is minimized. Valued metals are recovered from the leach effluent by essentially the same steps used in the hydrometallurgical portion of the smelting process.

Drawbacks to this process include, the care that must be taken in selecting materials for construction because of the high temperatures and acidities in the leaching step and the disposal of spent sulphate.

NOTES

1. Hillman, C.T., Manganese nodule resources of three areas in the northeast Pacific Ocean: With proposed mining — beneficiation systems and costs. A minerals availability system appraisal. US Dept. of the Interior, Bureau of Mines *Information Circular IC8983*.
2. Agarwal, J.C., H.E. Barner, N. Beechner, D.S. Davies, and R.N. Kust, Kennecott process for the recovery of copper, nickel, cobalt and molybdenum from ocean nodules, *Mining Engineering*, December 1979, pp. 1704–1707.
3. LIX64N is a trademark reagent of General Mills Chemical Company.
4. These characteristics were apparently successfully demonstrated in a 350 kg/day pilot plant which was operated for 43 days at Kennecott's Ledgemont laboratory. For a description of the above see Agarwal, J.C., H.E. Barner, N. Beecher, D.S. Davies, and R.N. Kust, Kennecott process for the recovery of copper, nickel, cobalt and molybdenum from ocean nodules, *op. cit.*
5. In respect of the alloy phase and its weight, reference is made to Monhemius, A.J., 1980, The Extractive Metallurgy of Deep Sea Manganese Nodules. Chapter in topics in nonferrous extractive metallurgy. *Soc. Chem. Ind.*, pp. 42-69.
   In respect of the technology see Dames and Moore, and EIC Corporation. *Description of Manganese Nodule Processing Activities for Environment Studies*, Vol. III. Processing Systems Technical Analysis (Contract 6-35331). United States Department of Commerce — NOAA, Office of Marine Minerals, Rockville, MD., 1977, 540pp.; NTIS PB274912.

6. *Ibid.*
7. *Ibid.*
8. Hillman, C.T., *op. cit.*

Chapter 7

Conceptual Flowsheets for First Generation Plants

MAJOR CONSIDERATIONS

Since manganese nodules represent a new kind of mineral resource, it is to be expected that the necessary process technology would involve some novel technological concepts. Due to the generally conservative nature of the minerals industry, however, it is unlikely that wholly new extraction concepts, spanning the treatment of raw ore to final metal production, would be adopted for a first generation nodule plant. The more likely scenario is one in which novel technology is judiciously grafted onto current knowledge and practice. In fact, the greater the component of conventional technology, the lower the risk factor and therefore, the better the prospects for adoption of a proposed integrated process scheme.

Much of the difference in the various proposed extraction schemes lies in the manner in which the destruction of the nodule matrix is effected.

The selection of a processing route by a developer would first be determined by the technical viability of the proposed technology. In addition, a number of external factors, such as those enumerated below, would influence the adoption of one among several technically viable options:

1. A unique situation regarding the developer's current plant practice (for example, plant location, raw material availability, technology, availability of partially depreciated equipment, product mix);
2. A unique position of the developer regarding the supply of process chemicals and reagents;

3. A unique position regarding the disposal of intermediate products or by-products;
4. Availability of power, water and other raw materials; and,
5. Local environmental protection and regulatory considerations and the cost of conforming with the regulations.

CONCEPTUAL FLOWSHEETS[1]

As previously noted, much of the difference in the proposed extraction schemes lies in the manner in which the destruction of the nodule matrix is to be effected. Based on an evaluation of the published literature by Dames and Moore, twelve potential routes to the recovery of the valued metals in manganese nodules using pyrometallurgical and hydrometallurgical processes, or combinations of the two, may be identified.[2] These routes ("roadmaps") or flowsheets, classified by the type of lixiviant used to solubilize the metals of interest, are summarized in Table 15 below.

Ammoniacal systems are used in processing land-based nickeliferous laterites which are, in some respects, similar to manganese nodules. Copper, nickel and cobalt are soluble in ammoniacal ammonium carbonate (the Caron process) and ammonium sulphate solutions. In these routes, by selectively reducing the metals of interest from their oxide states, the manganese nodule matrix is disrupted enabling rapid and complete dissolution of the valuable metals.

Hydrogen chloride solutions are also utilized in solubilizing the metal values of interest including manganese, and a substantial body of literature exists on such solutions. Copper, nickel and cobalt are also soluble in strong acid sulphate systems and serve as the initial step for various possible process routes.

The high-temperature H_2SO_4 leach process technology as used in recovering nickel from laterites, takes advantage of the fact that the high temperature increases the rate of dissolution of copper, nickel and cobalt, and limits the solubility of undesirable compounds such as iron and manganese. Alternative routes involving the acid sulphate lixiviant system include the selective, high-temperature reduction of the nodules, separation of manganese rich slags from the metallic phases, sulphidation of metallic phases and the subsequent selective leaching of the sulphide materials.

Table 15. Potential Process Routes (After Dames and Moore and the EIC Laboratories)

1. **Ammoniacal Systems**
   - (a)   Gas reduction and ammoniacal leach
   - (b)   Cuprion ammoniacal leach
   - (c)   High-temperature ammonia leach

2. **Chloride systems**
   - (a)   Reduction and HCl leach
   - (b)   Hydrogen chloride reduction roast and acid leach
   - (c)   Segregation roast
   - (d)   Molten salt chlorination

3. **Sulphate Systems**
   - (a)   High-temperature and high-pressure H_2SO_4 leach
   - (b)   Smelting and H_2SO_4 leach
   - (c) H_2SO_4 reduction leach
   - (d)   Reduction roast and H_2SO_4 leach
   - (e)   Sulphation roast

Of the twelve generic process types presented in Table 15, seven are known to have sufficient technical problems to preclude the likelihood of their commercial development. The five remaining process types are considered the best options at the present time for first generation nodule processing plants. Flowsheets for these five options have been developed both from the published literature and by analogy to the processing of land-based ores. The five process options are:

1. Gas reduction and ammoniacal leach;
2. Cuprion ammoniacal leach;
3. High-temperature and high-pressure H_2SO_4 acid leach;
4. Reduction and HCl leach;
5. Smelting and H_2SO_4 leach.

PROCESSES FOR FIRST-GENERATION NODULE PLANTS

The processes for recovering the valuable metals from nodules can be subdivided into those processes whose objective is to recover three metals (nickel, copper and cobalt), otherwise known in the literature as three-metal systems, and those processes that are intended to recover four metals (including manganese). The basic three-metal systems are:

1. gas reduction and ammoniacal leach process;
2. cuprion ammoniacal leach process, and,
3. high-temperature and high-pressure H_2SO_4 leach process.

The other two processes are considered four-metal processes.

The recovery of manganese from the tailings of the three-metal systems varies according to the lixiviant type used. In the case of tailings from ammoniacal leach, manganese can be recovered to a limited extent by the flotation of manganese carbonate ($MnCO_3$). The $MnCO_3$ could then be further processed using conventional technology to produce a manganese oxide product. The manganese oxide product could either be sold as such or further processed to produce ferromanganese or other manganese alloys.

In the case of tailings generated from the sulphuric acid systems, they can be chemically manipulated to recover the manganese as an oxide and, again, if desired, further processed to produce ferromanganese or other manganese alloys.

FIRST-GENERATION NODULE PLANTS: CONSIDERATIONS IN RESPECT OF LAND-BASED PRACTICES

Of the two ammoniacal leach schemes, the gaseous reduction route has the advantage of being similar to commercial laterite processing practices at Nicaro in Cuba, Surigao in the Philippines and Yabulu in Australia. However, it should be noted that laterite processing technology is itself undergoing changes. When the first reductive roast-ammonia leach plant was built in Nicaro, it was a one-metal plant. There was no nickel-cobalt separation and a market product of nickel oxide sinter was produced which analyzed 88% nickel and 0.7% cobalt. On the other hand, the two newer plants commissioned in Surigao and Australia are two-metal plants in that the products for marketing consist of a nickel product (nickel briquets in Surigao and nickel oxide sinter in Yabulu) and a mixed nickel-cobalt sulphide precipitate which is sold for further processing elsewhere. In fact it is conceivable that given the recent advances in solvent extraction technology, a new ammoniacal laterite leach plant would involve an integrated process at one location that takes the raw ore through leaching and solvent extraction operations to separate metallic products of nickel and cobalt. Research and development work (up to the stage of a pilot plant) by Universal Oil Products, Inc. (UOP) and the US Bureau of Mines has demonstrated the feasibility of such an approach.[3]

When comparing the processing of manganese nodules and laterites, the fact that the former are mined as wet ore containing up to 40% water (up to 30% water when air dried) would put a heavy burden on the energy costs of the reductive roast operation. In fact, even in the case of laterites this high energy operation represents a major item of cost and recently both the Surigao and Yabulu plants have been forced to switch from oil-based energy plants to coal. Thus, even though the cuprion ammoniacal leach process involves a novel and as yet non commercialized leaching technique, its ability to permit the direct processing of wet ores makes it very attractive. Moreover, it has been extensively tested at both bench scale and pilot plant scales by Kennecott. In the pilot plant test, the ore feed rate was 350 kilograms per day over a 43 day period of which 20 days involved a continuous run.

The high-temperature sulphuric acid pressure leach process is based on a similar process currently applied to lateritic ores at Moa Bay, Cuba. Thus, like the reductive roast–ammoniacal leach process, this processing scheme has the advantage of a high degree of prior technological knowhow. However, here again, the analogy with current laterite processing practice must take into account the fact that a new laterite facility based on high temperature sulphuric acid pressure leaching would not necessarily be a blue print of the Moa Bay plant. As an example, AMAX has introduced extensive modifications in the laterite leaching process as well as in the solution purification and metal production operations.[4]

The reductive chloridization–hydrochloric acid leach is a novel process scheme whose major components have no current commercial counterparts. It has the advantage of high recoveries of copper, nickel, cobalt and manganese. However, the need to dissolve all the manganese with an expensive reagent such as hydrochloric acid is a major disadvantage. Unlike the above schemes which are three-metal processes, the reductive chloridization–hydrochloric acid leach process is a four-metal process.

The matte smelting-sulphuric acid leach process has the advantage that it is currently practiced in the laterite industry. Matte smelting and leaching is also an established technology in the processing of copper, nickel, and cobalt sulphide ores. The major drawback is the high energy cost associated with the high temperature treatment of wet ore. However, as Sridhar *et al.* have argued, with efficient pyrometallurgical design (e.g. with emphasis on the use of fossil fuel rather than electrical energy), full advantage can be taken of this process which permits the early concentration of the valuable metals into a small volume of intermediate product, i.e., matte (with a resulting decrease in the size of the hydrometallurgical plant) while at the same time rejecting manganese as a slag product which can be readily stockpiled for future treatment if necessary.[5]

FIRST-GENERATION NODULE PLANTS: SUMMARY PROCESS DESCRIPTIONS

As mentioned in the preceding sections, of the twelve generic process types presented in Table 15, only five process types are considered as possible options at the present time for first generation nodule processing plants. Brief descriptions of each of the five processes follow.

Gas reduction and ammoniacal leach process

This process involves carbon monoxide gas reduction followed by an ammoniacal leach leading to the recovery of copper, nickel and cobalt. A simplified flowsheet for this process is shown in Fig. 12.

Manganese dioxide, the major constituent of nodules is reduced to manganese oxide by a carbon monoxide-rich producer gas at a temperature of 625°C. As a result of this reduction, the mineral structure of the nodules is disrupted to release the contained metals. The metal values, when solubilized, are dissolved from the reduced nodules with a strong aqueous solution of ammonia (10 percent) and carbon dioxide (5 percent), at low temperature (40°C) and atmospheric pressure.

The metal-bearing solution is decanted from the nodules and treated with a series of organic extraction steps, which selectively remove the copper and nickel from the aqueous solution. The metal values are selectively stripped from the organic extract

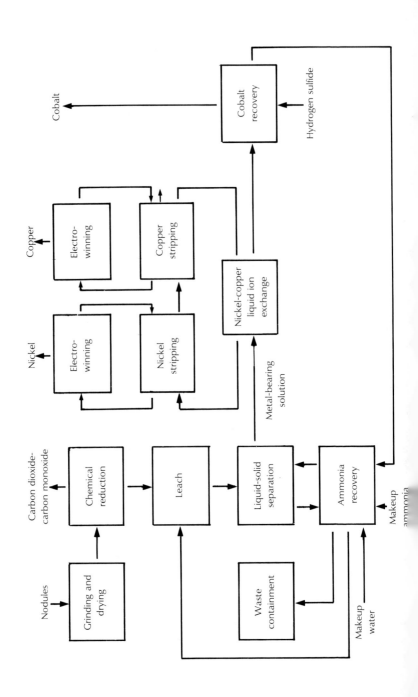

with acidified aqueous solutions. The metal products, copper and nickel, are produced from these acidic solutions by electrowinning. By contacting the aqueous ammonia-carbon dioxide solution with hydrogen sulphide, the insoluble sulphides of cobalt as well as small amounts of copper, nickel, zinc and other metals not removed in previous steps are precipitated. The undissolved sulphides are sold as minor products and the cobalt and nickel are recovered from solution in powder form by selective reduction with hydrogen at high pressure (34 atmospheres) and temperature (185°C). The nodule residue, from which the major portion (98 percent) of the soluble metals has been removed, is contacted with steam (at 120°C temperature and 2 atmospheres pressure) to remove residual ammonia and carbon dioxide.

Plant services for this process will include facilities for generating the producer gas used in nodule reduction, raising the necessary steam and part of the power required for process use, supplying the makeup and cooling water required, and providing for materials handling for process materials and supplies. The generation of producer gases from coal or oil for the reduction of nodules and all other plant services represent the utilization of known technology.

Cuprion Ammoniacal Leach Process

Copper, nickel and cobalt can be recovered from nodules from this process for which a simplified flowsheet (Fig. 13) has been presented. Using an aqueous ammoniacal solution containing an excess of cuprous ions (Cu^+) and at a temperature of 50°C, the manganese dioxide (MnO_2) in nodules is reduced to manganese oxide (MnO). The metal values released in this first step are then solubilized with a strong aqueous solution of ammonia and carbon dioxide at low temperature and pressure. The metal-bearing solution is then decanted from the nodules and, through a series of organic extraction steps, copper and nickel are then selectively removed from the aqueous solution. The organic extract, containing the metal values, is then treated with acidified aqueous solution from which cathode copper and nickel are produced through electrowinning.

When the remaining aqueous ammonia-carbon dioxide solution is treated with hydrogen sulphide (H_2S) cobalt, as well any residual copper, nickel and zinc in solution are precipitated. Upon treatment with air and hot (100°C) sulphuric acid, cobalt is

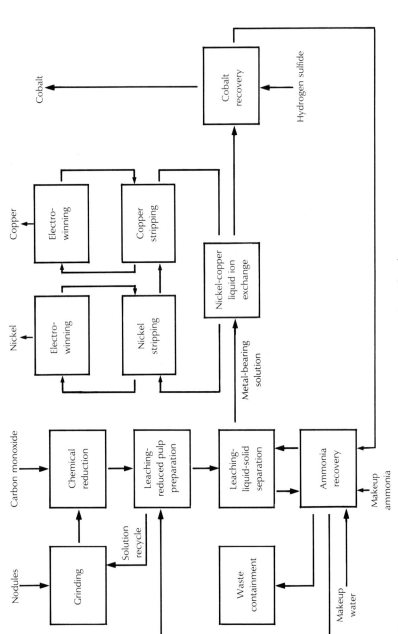

Fig. 12 Cuprion ammoniacal leach process

selectively redissolved and removed from the solids (precipitates). The undissolved sulphide precipitates can be sold as minor products while the cobalt and nickel are recovered from solution in powder form.

The nodule residue from which ninety-eight percent of the metal values have been removed, is then contacted with steam (at 120°C temperature and 2 atmospheres pressure) to remove residual ammonia and carbon dioxide.

Plant services for this process include facilities for generating the carbon monoxide used in nodule reduction, raising the necessary steam and power required for process use, supplying the makeup and cooling water required, and providing for materials handling for process materials and supply. The generation of carbon monoxide gas from coal or oil for the reduction of cupric ion (Cu^{2+}) and all other plant services represent the utilization of known technology.

High-Temperature and High-Pressure Sulphuric Acid Leach Process

The first step in this process is a high-temperature (245°C) and high-pressure (35 atmospheres) treatment of the ground nodules. A simplified flowsheet for the process is presented in Fig. 14. With the exception of manganese, most of the major metal values in the nodules become dissolved in the hot, strong (30 percent) sulphuric acid solution. After cooling, the nodule residue and acid solution are separated by decantation. Water is used to wash the residue free of acid and soluble metals.

Before the extraction of copper and nickel, the metal-bearing acid solution passes through a pH-adjustment step. Copper and nickel are then removed from solution with an organic extractant. The extracted nickel and copper are separately and selectively stripped from their respective organic extracts and transferred to acidified aqueous solutions, which accumulate nickel and copper sulphate. The metal products, cathode nickel and copper, are produced from these acidic solutions by electrowinning.

Cobalt is subsequently recoverd by precipitating the solution with hydrogen sulphide which also precipitates residual copper and nickel. The solid residue is recovered from solution and contacted with air and hot (100°C) sulphuric acid to selectively redissolve the cobalt and the small amount of nickel present. Again, the undissolved sulphides may be sold as minor products while the cobalt and nickel can be recovered in powder by

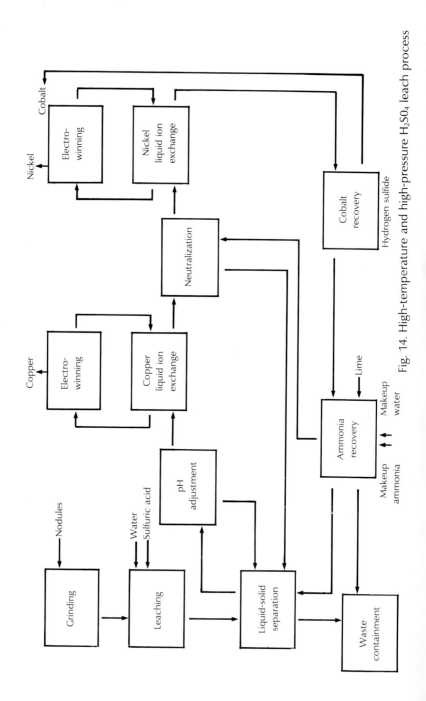

Fig. 14. High-temperature and high-pressure H_2SO_4 leach process

selective metal reduction through the use of hydrogen gas at high pressure (34 atmospheres) and temperature (195°C).

Plant services include facilities for generating the necessary steam and part of the power required for process use, for supplying the makeup and cooling water required and for providing materials handling for process materials and supplies. The generation of steam and all other plant services represent the utilization of known technology.

Reduction and Hydrochloric Acid Leach Process

Copper, nickel, cobalt and manganese can be recovered from nodules by this process for which a simplified flowsheet has been presented in Fig. 15. The basis of the process is the reduction of the manganese dioxide nodule matrix with hydrogen chloride to yield soluble manganese chloride, thereby releasing the nickel, copper and cobalt for dissolution.

The reduction and HCl leach process is a four-metal process in which manganese, copper, nickel and cobalt are liberated from dried nodules by a high-temperature (500°C) gaseous hydrogen chloride treatment of nodules. Hydrogen chloride reduces manganese dioxide to manganous chloride (liberating chlorine gas) and also reacts with other metal oxides to form soluble chloride salts. A hydrolysis reaction and quench follow, where water is sprayed on the nodules and the iron is precipitated as ferric hydroxide. The nodules are leached with water and hydrogen chloride, forming a concentrated pregnant liquor of chloride salts.

Copper is extracted by liquid ion exchange (LIX) reagents from the pregnant liquor, and is stripped and recovered as electrowon cathodes. Cobalt is extracted from the copper raffinate, stripped, and separated by precipitation with hydrogen sulphide. It is recovered from the sulphide precipitate, along with some nickel, zinc and copper, by selective leaching and hydrogen reduction. Nickel is extracted by (LIX) reagents from the cobalt raffinate, stripped and recovered as electrowon cathodes The nickel raffinate is evaporated, crystallizing manganese chloride as well as other remaining chloride salts.

The salts are dried using combustion gases in a countercurrent dryer. The dried salts are charged to a high-temperature fused salts electrolysis furnace, where molten manganese metal is tapped and cast as product and chlorine gas is liberated. Excess hydrogen chloride gas in the process is recovered and recycled.

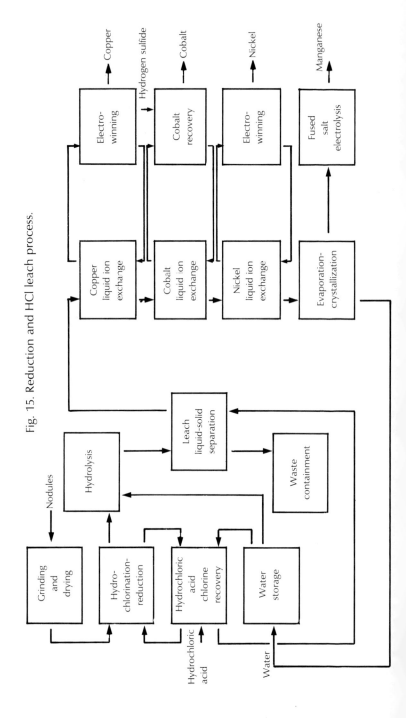

Fig. 15. Reduction and HCl leach process.

Generated chlorine gas is recovered, dried, and delivered to a local chemical complex which, in exchange, returns makeup hydrogen chloride to the process.

Plant services include process and cooling water supply and treatment, steam raising and power generation, stack gas treatment and combustion gas preparation. Makeup water is clarified and softened for distribution to the process, as required. Additional treatment is required for cooling tower water makeup, boiler feed makeup and for supplying plant potable water. Process combustible wastes are burned along with coal in the wain boilers to raise the required process steam and generate a portion of the power required in the process.

Smelting and Sulphuric Acid Leach Process

The smelting and sulphuric acid leach process is a combination of pyrometallurgical and hydrometallurgical treatment of nodules to recover copper, nickel and cobalt with the option of recovering ferromanganese or a storable byproduct of manganese and iron. A simplified flowsheet of this process is presented in Fig. 16. The smelting process produces a slag, from which ferromanganese is recovered, and a metal alloy matte composed primarily of nickel, copper, cobalt and sulphur.

The matte is granulated, slurried and selectively leached with sulphuric acid at elevated temperature and pressure. The leach residue and metalliferous solution are separated by a series of filtering and washing stages. After liquid-solid separation, copper and nickel are selectively extracted by liquid ion exchange (LIX) reagents, stripped from the ion exchange liquid into a weak electrolyte and recovered as electrowon cathodes. Cobalt is separated from the raffinate by precipitation with hydrogen sulphide and recovered from the sulphide precipitate, along with some nickel, copper and zinc by selective leaching and hydrogen reduction. Ammonia consumed in the process is recovered by lime boil and recycled to the process for use in pH control.

Plant services include process and cooling water supply and treatment, steam raising and power generation, stack gas treatment and producer and dryer gas production. Makeup water is clarified and softened for distribution to the process as required. Additional treatment is required for cooling tower water makeup, boiler feed water makeup and supplying plant potable water. Offgas hydrogen from cobalt recovery and offgases from the

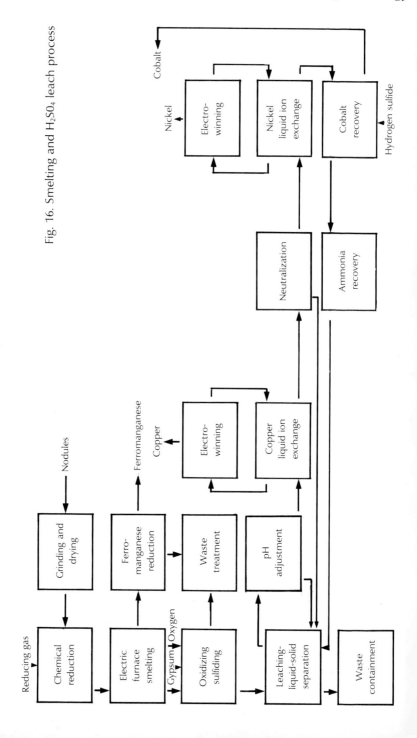

Fig. 16. Smelting and H$_2$SO$_4$ leach process

reduction and smelting steps are burned in the main boiler, along with coal, to raise the required process steam and generate a portion of the power required in the process. Following particulate removal, the gases are combined with other process offgases and are sent for gas treatment, where sulphur oxides and other acidic constituents are removed by scrubbing with limestone. The scrubbed offgases are reheated, combined with scrubbed vents from various process steps and sent on to the stacks for disposal.

NOTES

1. Haynes, B.W., S.L. Law and R. Maeda, 1983. Updated Process Flowsheets for Manganese Nodule Processing, United States Bureau of Mines *Information Circular IC 8924.*
2. Dames and Moore and EIC Corporation, 1977. *Description of manganese nodule processing activities for environmental studies,* Vol. III. Processing Systems Technical Analysis, (Contract 6-35331) US Dept of Commerce — NOAA, Office of Marine Minerals, Rockville, Md., 1977, 540 pp., NTIS PB274915.
3. (i) Stevens, L.G., L.A. Goeller and M. Miller, 1975. The UOP Nickel Extraction Process. An improvement in the Extraction of Nickel from Laterites, 14th Annual Conference of Metallurgists, The Canadian Institute of Mining and Metallurgy, Edmonton.
   (ii) Nilsen, D.N., R.E. Siemens and S.C. Rhoads, 1980. *Solvent Extraction of Cobalt from Laterite-Ammoniacal Leach liquors,* US Bureau of Mines Report of Investigations 8419.
4. Duyveskyn, W.P.C., G.R. Wicker and R.E. Doane, 1979. An omnivorous Process for Laterite deposits, in *International Laterite Symposium,* pp. 553–570.
5. Sridhar, R., W.W. Jones and J.S. Warner, Extraction of Copper, Nickel and Cobalt from Sea Nodules, *Journal of Metals,* April 1976, pp. 32–37.

Chapter 8

Conclusions

Based on a thorough review of the technical and patent literature on the processing technology for manganese nodules that has been developed over the last decade, it can be concluded that such technology is well developed on a laboratory scale basis. Although manganese nodules are a very complex ore, the basic processing routes being considered are, for the most part, extensions of ones used for other types of ores, e.g. nickeliferous laterites. At certain stages of the various processing routes, given the differences in the chemical composition of nodules and similar land-based ores, there is uncertainty about how some conditions necessary for desirable reactions are brought about, and these stages are the subject of proprietary research which has led and will continue to lead to the granting of patents. However, since nodules have not been processed beyond a bench scale, it is to be expected that many significant modifications to the original technical configurations will have to be made before their processing can be carried out at a commercial scale.

At the present time, the processes considered most feasible are based on smelting or on leaching with hydrochloric acid, sulphuric acid, or ammonia. Nodules processing bears a great similarity to nickel laterites processing, therefore energy costs and capital charges would be expected to be about the same. Thus, other costs, particularly for process materials, would need careful control to be cost-effective. The processing route chosen must, therefore, be closed in respect of the lixiviant used or a special situation must exist for adding or disposing of spent lixiviants.

Among the candidate routes proposed, low-temperature sulphuric acid processes as well as segregation roasting show rather low yields even with long leach times and as result do not appear to be the routes likely to be selected. High-temperature sulphuric acid leaches show good recoveries in shorter time periods while leaches based on ammonia or ammonium carbonate, also well known in the processing of laterites, produce excellent metal recoveries and are very selective.

When manganese recovery is considered, the smelting route will be a standard route for the production of ferromanganese. The sulphating roast and sulphuric dioxide reduction leach processes are candidates for manganese recovery since metal solubilization is high. There are problems, however, in integrated processes based on the above. In the case of processes based on sulphuric acid, the acid produced as the metals are recovered from solution is unlikely to be economically recyclable and its disposal will require neutralization. This will necessitate the use of large quantities of bases and result in the generation of large waste streams. The same considerations may be applicable to schemes where the lixiviant is ammonium sulphate and to some extent the hydrochloric acid based schemes.

The consensus in the literature is that the capacity of a nodules processing plant producing three metals, nickel, copper and cobalt, should be three million metric tons of dry nodules per year, with appropriate processes being based on sulphuric acid or ammoniacal leaching (Cuprion process).

The highest revenues from processing will be through the production of pure metals, and it is anticipated that all metals with the possible exception of manganese will be recovered in this form. Nickel and copper will be electrowon from solution to produce products of cathode specification. The metals will have to be separated from solution first for this to be done, and all the schemes involve liquid ion exchange (LIX), or chelating reagents. Cobalt will be produced as a hydrogen-reduced, briquetted powder. Minor amounts of nickel and mixed impure zinc and copper sulphides will be produced as a by-product of cobalt recovery operations. These sulphides may then be sold to smelters for further processing. Cobalt could be electrowon in some cases, such as in chloride leaching. Manganese will be produced as either standard ferromanganese from smelter slug and leach tailings or as electrolytic manganese metal by electrowinning from solution.

At the present time is impossible to determine when metal prices will rise to the point where interest in manganese nodules as a mineral resource will make for continued work in establishing the commercial viability of nodule processing. Looking ahead, however, the next step in nodule processing technological development will be production scale-up.

INDEX